Math Fun for Everyone

by

Werner Weingartner

Solves $1 + 1 = X$

iUniverse, Inc.
Bloomington

Math Fun for Everyone

iUniverse books may be ordered through booksellers or by contacting:

iUniverse
1663 Liberty Drive
Bloomington, IN 47403
www.iuniverse.com
1-800-Authors (1-800-288-4677)

ISBN: 978-1-4759-3614-8 (sc)
ISBN: 978-1-4759-3615-5 (e)

Printed in the United States of America

iUniverse rev. date: 7/2/2012

INTRODUCTION

Nobel laureate Eugene Wigner marveled at the "unreasonable effectiveness of math in formulating the laws of nature and even in predicting the world around us. Few branches of study gives man this God-like power."

This book is designed to be fun and a learning experience for those who appreciate the joy of math. I am a retired math teacher from the Bronx High School of Science and I know a thing or two about teaching math. The problems range from reasonable to somewhat difficult. I also decided to include little stories about how I developed my interest in math. You are in for an exciting adventure.

I have a neighbor April Cody, who is a 9th grade student in the honors math class and on the math team. She is a real nice girl and we both share an interest in math. April mentioned to me when she went to the library to take out a book in math she said most of them were intimidating, filled with strange symbols and concepts way over her head. She said, "Can't I just have fun with math and have a reasonable chance of solving the problems?" Well that's what this book is about. It is not designed for a PhD student in math and it isn't designed for the student who hates math and can't wait for math class to be over. It is designed for the good high school student and the millions of adults who just enjoy math.

Table of Contents

I have indicated the degree of math difficulty for each chapter, from 1=easiest to 5=hardest.

CHAPTER I

MY MATH HISTORY

Kindergarten 1941—We learned how to count from 1-100. I asked the teacher what is after 100? Does it ever end? At night I looked at the stars and wondered what's behind the star etc. etc. I was trying to understand infinity. A real shock came 12/7/1941, Pearl Harbor.

In 3rd grade I was very fortunate to have the greatest teacher in my learning career the one and only Miss Dancing. One day she asks the class, "does anyone know what 12×12 is?" I answered 144 and she said that was wonderful. It was one of the happiest moments of my life. She took me aside one day and told me a story of this brilliant kid, Karl Fred Gauss (he lived when George Washington was alive) who amazed the teacher with the following problem. His 3rd grade class had misbehaved and the students had to come in after school and add the numbers from 1 to 99. In order to show me what he did, Miss Dancing prepared me by looking at a simple problem: adding up the numbers from 1 to 9

$1+9=10=5+5$

$2+8=10=5+5$

$3+7=10=5+5$

$4+6=10=5+5$

$5=5=5$

We have now replaced all the numbers from 1-9 with a 5, therefore we have 9 5's= 9X5= 45

In other words, he changed an addition problem to a multiplication problem. Now we are prepared to add the numbers from 1-99.

1+2+3+4……96+97+98+99

1+99=100=50+50

2+98=100=50+50

3+97=100=50+50

4+96=50+50

He replaced each number by 50, therefore we have 99 50's=99X50= 4950. He changed a long addition problem into a multiplication problem. Obviously the teacher was amazed at the 3rd graders brilliance, wow!

Problems for the Reader: All solutions are at the end of each Chapter

Add all the numbers from 1-49

Add all the odd numbers from 1-99

Add all the even numbers from 1-99

Add the first 20 [5, 10, 15, 20, 25, 30…]

What is the 100th term?

Another problem Miss Dancing showed me was squaring numbers ending with 5.

Example:

$25\times25= 625$ $5\times5 = 25;\ 3\times2 = 6;\ = 625$

$35\times35= 1225$ $5\times5 = 25;\ 4\times3 = 12; = 1225$

$65\times65= 4225$ $5\times5 = 25;\ 7\times6 = 42; = 4225$

$85\times85= 7225$ $5\times5 = 25;\ 9\times8 = 72; = 7225$

Do you see what she did? I really enjoyed this problem.

A story comes to mind of a non math lesson she taught me. There was a girl, Roberta, who sat next to me that I had a crush on. I couldn't let any of the boys know about this because they would call me a sissy. One day I gave Roberta an ice cream cone that held two scoops. She was very impressed and mentioned to me that she is rooting for me to win the class president election. I was running against a very popular student and I knew I couldn't win. But I wanted to win this election and show Roberta what a big shot I was. I asked myself how can I win and I came up with an idea that I would give each student who voted for me a free comic book. Miss Dancing found out about my scheme and told me that this was wrong (she used the word bribery-I never heard that word before). Well if Miss Dancing said it was wrong then it was wrong. I took back the offer and lost the election. It was a bitter blow. Fifteen years later I met Roberta again and she was not anywhere as charming as she was in 3rd grade.

In 4th grade I had the mean spirited of Miss Fry. As much as I loved Miss Dancing, Miss Fry was no joy. One day she put the following problems on the board: $6 \div \frac{1}{2} = 6 \times 2 = 12$, $6 \div \frac{2}{3} = 6 \times \frac{3}{2} = 9$. I asked her what gives you the right to do that. She repeated invert and multiply. I asked her again why that does work. She was annoyed and said, "Just do it". For the next two hours I worked on this problem, I was thrilled when I uncovered the mystery. Teachers like Miss Fry should not be teaching, she was more suited to be a prison guard.

When I have 13½, I had the good luck to work with Irving Yano, who owned a very popular grocery store (this was before the big supermarkets). My job was to wait on customers, deliver orders and everyday have a math puzzle for him. Irving was a big husky man, a Russian immigrant who barely had an education in Russia but he loved math. Irving was a great story teller; he had an excellent sense of humor. His grocery store was so popular because of his compelling personality. This was a real fun job and I couldn't wait to get to my job after school.

I can see him now with a sugar cube in his mouth drinking tea. Every day he was eager to get his math problem and find out what I did in math and science.

Irving had several vices which were real eye openers for a 14 year old. Every day Al, the bookie, had a card game in the back of the store. It was common knowledge that Al sold untaxed liquor and had connections with some of the criminal elements. Al was 45 year old, a very likeable man but unbelievably he was a real momma's boy, who lived with his widowed mother. He jumped when mom gave an order. Irving, Al and Dr. Hess were all good friends, they told their family they were playing cards Wednesday nights, but they were not playing cards, they were involved in a different type of entertainment which didn't come up to standards of good family men.

One of Irving's favorite problems was the 12 coin problem. The problem goes like this. You have a balance scale and 12 coins. One of the coins had a different weight then the 11 good ones. In 3 balance scale you are to be about to find the bad coin. Irving told me only two of his friends could solve the problem I and was intensely pleased to receive Irving's congratulations when I solved it.

Can you solve the 12 coin problem? It is not easy but doable. *Hint*- you can move the weights from one side to the other side of the scale. Call the coins A_1, A_2, A_3, A_4, B_1, B_2, B_3, B_4, C_1, C_2, C_3, C_4 at weighing #1 weigh A_1, A_2, A_3, A_4 Vs. B_1, B_2, B_3, B_4

Another problem that Irving gave me and I solved was the four- 4's problem. Using four 4's write all the numbers from 1-22. Using +, -, ×, ÷, .4, √4. Note; you must use all four 4's.

For example: $1=\frac{4+4}{4+4}$ $11=\frac{4}{.4}+\frac{4}{4}$ $15=4\times4-\frac{4}{4}$ $21=4\times4+\frac{\sqrt{4}}{.4}$

$13=\frac{4+.4}{.4}+\sqrt{4}$ $19=\frac{4\sqrt{4}-.4}{.4}$

Irving was somewhat familiar with algebra and some of the laws of physics using the language of science, algebra. Irving stressed again and again, look for patterns. I was able to generate many problems using simple algebra. For example:

X	1	2	3	5	10
Y	2	4	6	10	20

$Y=2X$

X	1	2	3	5	10
Y	6	7	8	10	15

$Y=X+5$

X	1	2	3	5	10
Y	3	5	7	12	22

$Y=2X+1$

X	1	2	3	5	10
Y	1	4	9	25	100

$Y=X^2$

Then I gave more difficult patterns:

X	1	2	3	4	5	6	10
Y	0	2	6	12	20	30	90

$Y=X^2-X$

X	1	2	3	4	5	6	10
Y	1	8	27	64	125	216	1000

$Y=X^3$

X	1	2	3	4	5	6	7	10
Y	-1	1	-1	1	-1	1	-1	1

$Y=(-1)^X$

X	1	2	3	4	5	6
Y	2	-2	2	-2	2	-2

$Y=2(-1)^{x+1}$

I was able to come up with many combinations using X and Y. It often led to important scientific concepts. For example:

T	0	1	2	3	4	5	6	10
S	0	2	8	18	32	50	72	200

$S=2T^2$

It just so happens on the moon objects drop 2 feet in 1 second, 8 feet in 2 seconds, 18 feet in 3 seconds. In other words, we have a formula for falling objects on the moon.

Irving was quite taken with Einstein's $E= mc^2$. This simple equation shows the relationship of mass and energy. He would point to a crumb and say this little crumb has enormous energy. I pointed out a lot of things take on the form $y=kx^2$. For example, take Galileo's famous equation for falling objects on the planet earth; $s=16t^2$. In one second an object falls 16 feet. When t=2, s= 64 feet. When t=3, s= 144 feet.

T	0	1	2	3	4	5
S	0	16	64	144	256	400

Irving was familiar with the inverse relationship. Note as X gets bigger, y gets smaller.

$y= 24/x$

X	1	2	3	4	5
Y	24	12	8	6	24/5

But he was not familiar with the inverse square concept. This is a more difficult concept. As x is doubled, y is divided by 4, as X is tripled, y is divided by 9.

$y= 24/x^2$

X	1	2	3	4	5	10
Y	24	6	24/9	24/16	24/25	24/100

He was very excited when I showed him some of the most important ideas in science. The Law of Gravity: $G\frac{m_1 m_2}{r^2}$

In other words we can calculate the force of attraction between the earth and the sun, the earth and the moon, the sun and the moon. Both of us agreed how extraordinary it was that man can uncover the secrets of nature that follows such simple formulas. We had many philosophical discussions on how God let us in on his plan. As you can see this was more than just a fun job. God let man into the inner sanctuary of his/her domain. E = mc^2, C= 186,000 miles a second and

[186,000]^2 implied that even his cube of sugar had so much energy. We both agreed it was almost as if man had a direct connection with God. There is order in the universe and man has a chance to uncover these secrets. At the time I had the most philosophical discussions I ever had. Irving stressed again and again to look for patterns. This idea really stuck with me.

Newtons equation for gravity had a profound effect on me at age 14. Sixty years later I wrote the following memorable Valentine's card to my wife on February 14, 2010:

Dear Mary Anne,

There are two great mysteries of life: gravity and love. We know how each feels bu we can't explain how they work. For gravity, there is a formula which shows the relationship between the apple and the stars. For love, there is no such formula, but I an in your gravitational field which I feel every day in every way. Love, Werner

When I was 14½ years old, there was an incident in the store which I will never forget. A very nice man, Fred, used to deliver supplies to the store every Wednesday. One Wednesday, there was a substitute for Fred who made deliveries. None of us knew that this was his 2nd week out from prison and he was not making a good adjustment to his unfamiliar surroundings. He came in with his hook moving boxes around and showing complete disregard for Irving's costumers. Irving took exception and told this man not to mistreat his customers. A big argument developed and matters were clearly getting closer and closer to violence. The man threatened him with his hook and Irving who never backed off from a fight broke a large bottle of beer exposing the jagged edges. I had to do something and I was scared out of my wits. I stepped in between the two men and said, "Gentlemen I am sure this problem can be solved." My peacemaking effort had been ignored and the men were cursing and threatening each other. It was clear to everyone this was a very dangerous situation. Finally I said today is Sunday and you men should not be fighting on a Sunday. This broke the ice because it was a Wednesday and not a Sunday, and it was clear I was beside myself to keep the peace. I told the man I would bring in the delivery and everyone should calm down. Irving had enough sense to back off and let me take in the delivery without a hook. I know Irving was grateful to me for de-escalating this life threatening situation.

In 1952 Irving informed me that in his opinion the neighborhood was changing and a little grocery store would be replaced by supermarkets. He bought a supermarket in an up and coming neighborhood and he prospered financially but unfortunately the story does not have a happy ending. He had a son Lester who caused a lot of trouble in school. Teachers sent notes home about his unacceptable behavior in school. I urged Irving to give this matter careful consideration and discipline his son. It was Irving's opinion that's the way boys behave. In 1955, Lester got in big trouble getting involved in an armed robbery in Athens, Georgia. Irving was heartbroken and in four months he died of a heart attack. Irving treated me like a son and I lost one of my big heroes.

I had a lot of interesting different jobs which were more fun than school. Unfortunately I didn't have much time for homework and it negatively impacted my grades. One of my little jobs was working for Mr. Eggenhaus who had a linoleum and rug business. One of my jobs was to take measurements in people's homes to determine how many square yards were needed. One very memorable experience was when Mr. Eggenhaus sold a very large roll of linoleum (which had already been partially cut). He didn't know its area so he considered rolling the piece down Broadway to get the measurement. I told him I could calculate the area without unrolling this huge piece. He said that was impossible but I assured him I could do it. Here is what I did: I found the outside and inside radius, took the average, multiplied the average by the radius × height 6' × 2π R [the number of rings]÷ 9 and came up with the answer. He asked me to double check and I told him I was positive. He thought I was a genius when in reality it was not a hard problem.

At 15 another one of my jobs was to tutor math where I was paid $3.50 an hour. I tutored Susan, Dr. Hess's daughter. I was given the task to help Susan pass the geometry regents. She was not a bad student but she really disliked math and did as little work as possible. I tried very hard to direct her to her studies, but she had other ideas. In the middle of the lesson she wanted to practice her romantic skills with me. Net result, we spent half our lessons on things other than math. I felt a little guilty of accepting $3.50 an hour for these romantic adventures.

A real tough case was the Prystowsky family. I was very friendly with Seymour and he always complained how his mother was always on his back about school. The Prystowskys were a Russian immigrant family and Mrs. Prystowsky wanted in the worst way to set her High School Diploma. One thing stood in her way she had to pass the geometry regency. She took the regency 2 years ago and she got an 18. The next year she took it again and she got a 24. I had the reputation in the neighborhood as being a miracle worker in helping students pass math. She told me she was desperate and she will put herself in my hands. She was a very hard working highly motivated; her favorite expression was "Education comes first in this house". Unfortunately she had lost complete confidence in her math skills and she had unbelievably little talent in math. She asked me, "Do you think you can do something with me." I lied to her and said yes I can get her to pass the exam. This was a real challenge. I designed a plan for her who was to concentrate on calculation and bypass the formal proofs. To get her confidence back I gave very easy problems which she solved. It was my plan to give her feeling that she had a real chance of passing.

Euclid would not have approved of my method but Mrs. Prystowsky was thrilled to get a 72 on the regency and she got her High School Diploma. Her son Seymour complained a lot about his oppressive mother, but Seymour is now a very successful doctor.

Answers Chapter I Tests

1 to 49

$1 + 2\ldots48 + 49$

replace each number by 25

$25\times49 = 1225$

Sum of the odd numbers to 100

$1 + 3 + 5 + 7\ldots97 + 99$

replace each number by 50 (there are 50 numbers) $50\times50 = 2500$

Sum of the even numbers to 100

$2 + 4 + 6\ldots98$

replace each number by 50 (there are 49 even numbers)

$50\times49 = 2450$

Sum of all numbers divisible by 5 to 100

5, 10…95, 100

replace each number by 50.5

$(50.5)(20) = 1010$

500

12 coin problem $A_1, A_2, A_3, A_4, B_1, B_2, B_3, B_4, C_1, C_2, C_3, C_4$

Weighing I: A_1, A_2, A_3, A_4 Vs. B_1, B_2, B_3, B_4 if they balance then one of the C's is odd coin. If a side goes down, then either one of A's is heavy or one of B's is light. If a side goes down place A_3 B_4 aside.

Weighing II: A_1, B_2, S_8 Vs. A2, A_4, C_1

You should now be able to finish.

4-4 Problem

At the end of each chapter, I would like to present a little surprise that has little to do with math.

Here are three of my favorite works of art.

The Luncheon of the Boat Party By Renoir

2. The Great Wave Off Kanagawa

By Hokusai (1831)

3. Potato Eaters By Van Gogh

CHAPTER II

FOR THOSE WHO HATE MATH

Do math and computers make you unhappy?

YOU ARE NOT ALONE

In 2008, I wrote this book called Math Fun for Everyone. It was filled with exciting problems, puzzles, and aspects of math which I thought would appeal to a broad base of the population. My wife looked over my first try at this book and told me if I was serious about having this book accepted, it would have to be more appealing to a broader base than just math nerds. I have learned again and again to listen to my wife. I have lightened up on some of the math and mixed in stories that are tangentially involved with math.

The net change has markedly improved this book.

In my opinion, Chapter Two will prove to be one of the all-time most exciting and adventurous chapters in your reading history. I will start with ten little stories.

November 10, 1938

I was a two-and-a-half year old living in Germany when Crystal Night occurred. All over Germany, Jewish shops and synagogues were set on fire. Germany was a very dangerous place for Jewish people to be. On November 11, 1938, Kate Smith introduced and sang Irving Berlin's "God Bless America," for the first time.

In February 1939, this blessed country saved our lives by opening its doors to us. This one short song so captures my feelings for America.

While the storm clouds gather far across the sea,
Let us swear allegiance to a land that's free,
Let us all be grateful for a land so fair,
As we raise our voices in a solemn prayer

God Bless America,
Land that I love,
Stand beside her and guide her
Thru the night with a light from above;

From the mountains, to the prairies,
To the oceans white with foam,
God bless America,
My home, sweet home.
God bless America,
My home, sweet home.

I was so moved by this song that in high school, I would argue with English teachers that the lyrics of this song is poetry of the highest order. I still feel that the Gettysburg Address, Declaration of Independence contain poetry that far surpasses the poetry that so many English teachers cherish.

Mr. Eggenhaus story

On July 3, 1951, three interesting things happened to me within a two day period. Mr. Eggenhaus, who owned a linoleum and rug store, presented me with a problem. He had a big roll of linoleum and it was important for him to find the area. He asked me if it was possible to calculate the area without having to roll open the piece down Broadway.

I told him I was fairly certain he could calculate the area without having to open it up. Here is what I did. The inside radius of the linoleum roll was two inches, and the outside radius was ten inches, and there were 36 rings in the roll. The average radius was six inches (0.5 feet). The area of the linoleum roll was $2\times\pi\times0.5$, which I then multiplied by the height of the roll (6 feet), and then again by 36 to account for each ring. I divided by 9 to determine the area in square yards, and found out that this linoleum roll had an area of approximately 75 square yards. Mr. Eggenhaus thought I was a genius but in reality, it wasn't a very hard problem.

Mr. Eggenhaus was a very nice man of simple wants. He constantly tried to improve his store by broadening the variety of things he could sell. He was also very eager to marry off his not-so-beautiful and not-so-charming daughter Schler. He must have known she was no bargain but had plans to take a future son-in-law into the business. It was like the old dowry system, which provided an encouragement for a young man to marry a girl. His plan worked very well, because he found a man for his daughter and successfully expanded his store to include closeout furniture and those with minor defects. This was another American success story.

As a teenager, I had two problems I struggled with: I didn't know how to swim, and I had a lisp. I couldn't say words with "th" in them, like "Ruth" or "through." On July 4, 1951, I was in a public swimming pool and there was lightening overhead. everyone left the pool but I was convinced I would be able to swim. Yes, I did swim and the next day my lisp went away, never to return. Unfortunately, I was not so successful in solving all my teenage problems so easily.

My first day teaching

I was the fifth teacher for this unruly math class in New York City in January 1958. It was my first day teaching, and the students were running around completely out of control. It was clear to me I was in big trouble. I took a yard stick, hit the desk, and said "This is a direct order. Every student is to take their seat." It was clear to everyone that something dramatic would happen. I felt like I was a sheriff in an old Western movie. Every student but one cooperated. I repeated my order and he said "FU." The sheriff had to act and so I grabbed the kid and hit his behind again and again. It may not have been within Board of Education regulations, but it was the right thing to do. It was a good solution to a difficult problem. Yes, I did survive and in time proved to be an outstanding math teacher.

Irene Mills

In October 1965, I called on Irene Mills, one of the best students in my geometry class and asked which postulate I was using in a problem. It was clear she didn't know the answer, but came up with the following unforgettable gem: "The postulate of common sense that overrules all other postulates." WOW!

On April 1, 1967, I was teaching an Honors 12th grade math class and the students wanted to make an April Fools' joke. They enlisted Irene Mills, one of my favorite students, to set up the joke. Irene asked, "Mr. Weingartner, we are all very impressed with your knowledge of math, but the class would like to ask you what does 1+1 equal?" The class laughed and called out "April Fools!"

I surprised them with my unexpected answer. I mentioned

1+1=0 (mod 2)

1+1=2 (Boolean algebra)

1+1=10 (base 2)

Then I added that if a boy and girl are properly inclined, 1+1=3. They all laughed and Irene said she was thrilled and shocked by my answer. As a result, I almost called this book "1+1=X".

Four years after she graduated from high school, Irene became a math teacher. In eight years, she became a principal. Irene was a real treat.

The Job Interview

In 1979, I was very eager to get a math position at the Bronx High School of Science. I took a day off and went to the school hoping to make a good impression, which would lead to a teaching position in the famous school. When I entered the building, I saw a man who I thought was Bob Weinberger, a colleague from the past. I went over to him, slapped him hard on the back and said, "Bob,, what a pleasure to meet you here." The man looked shocked and told me not to slap him on the back again and asked who I was and what I wanted. I told him I was there to see the principal about a math position in the school. He told me he was the principal and that I sure hadn't made a very good first impression. What I embarrassed! By the way, I did get the job and had many enjoyable years teaching at the school.

Sexual Problems of Jewish Men

In 1983, the famous sexologist Dr. Ruth Westheimer sent out flyers about her talk "The Sexual Problems of Jewish Men." Of course I attended and while she was giving her talk, I was preparing my lesson for my Advanced Placement calculus class by making up notes on the back of the flyers. The next day I am giving my lessons in the Advanced Placement calculus class when a girl noticed the flyers that said, "The Sexual Problems of Jewish Men." The class was hysterical with laughter and I was beet red.

The Power of Music

I used to love to play paddleball, and in January of 1976 after playing several games, I came home and opened the windows, even though it was cold. I put on this record of Richard Tucker singing Puccini and took a warm bath. I found this experience thrilling, and was filled with a sense of enormous joy and happiness. The mixture of the music, warm water and the cold air coming into my little apartment put me in a euphoric state. After the bath, I sat down and wrote a letter to Richard Tucker expressing my delight with his recording and sent it in care of the record company that produced it. Three weeks later I got a letter from Richard Tucker's widow informing me that Richard Tucker died six months previously, but she was sure he would have loved my note. It was somewhat embarrassing.

In 1979, I went to hear a Russian violinist making his debut performance playing Tchaikovsky's Violin Concerto. It is one of my favorite pieces and I know it is a very difficult piece to play. I sat in the audience and I was enthralled with the performance. It was marvelous. then it came to what I thought was the end of the piece--I was so excited by the performance that I got up and yelled "Bravo, bravo!" But I was the only one cheering because I didn't realize there was another two minutes to go. I was very embarrassed standing up alone cheering before the end of the piece.

The Last Major Conversation with my Father

On June 1, 1989, my 88-year-old father got terrible medical news. When I spoke with my father, I wanted to determine if he had all his faculties after this terrible news. I asked him was year I was born, and he answered 1836. I pointed out that I was born in 1936, not 1836. His answer was "You wanted to see if your father was all there." Then he said, "Look, Werner, I am not so afraid to die because I will be with your mother and I will be very happy. Does this answer your question?" WOW.

The Cheerios story

This took place two days after Valentine's Day in the year 2000. On Hilton Head Island, where I lived, the local store had a sale on Cheerios for $1.59 a box. I like Cheerios, so I went to the store with six Cheerios coupons that I intended to use in conjunction with the $1.59 cereal ad. When I got to the register with my six boxes of Cheerios, the cashier rang them up as $3.59, not $1.59. I raised my voice and told her its $1.59. The manager came over and informed me that the sale is for Honey Nut Cheerios. I got annoyed and again raised my voice and the cashier began to cry. Then I noticed a sign that stated the cashier was hearing impaired. I felt terrible. How in the world could I make this deaf girl cry over a box of Cheerios? I felt awful and I came home and told my wife what had happened. She agreed this was embarrassing. That night, I couldn't sleep thinking over what I had done. To make matters worse, I kept up my wife who got very annoyed with me. Now I had two women mad at me.

The next morning, I got up and I came up with a solution to my problem. I went back to the store and bought flowers for the cashier and for my wife. On my wife's note, I wrote, "Valentine's Day has come and gone but my love for you lingers on and on." What a great solution!

On February 14, 2010, I wrote a Valentine's Day card that I am very proud of.

Dear Mary Anne,

There are two great mysteries of life: gravity and love. we know how each feels but we can't explain how they work. For gravity, there is a formula which shows the relationship between the apple and the stars. For love, there is no such formula, but I am in your gravitational field of love which I feel every day in every way. Love, Werner

I am really proud of that Valentine's Day message.

2010 and Beyond

In 2006, I decided to make a 92 minute movie called "1+1=X." I thought it was a great movie but the rest of the world didn't share my opinion. If you want to see a trailer, go to Google and search "movie one plus one = x."

I am going to try and edit this movie to make it a commercial success. Any and all profits will go to the Weingartner Global Initiative at the College of William and Mary. The probability of success is low.

I am going to launch my book in 2010. It is very hard to get a book published, so the chances of success are less than 50-50.

I am very involved with the crown jewel of my life's work called The Weingartner Global Initiative at the College of William and Mary. I think this program has an excellent chance at succeeding.

Improving Math Education in America

1958 was my first year of teaching and it was mandatory to take these silly education classes to qualify to teach math. At the time, I learned a new word ("compunction") that I was eager to use. I went over to one of my education professors and asked him if he had any compunction in teaching this nonsense year in and year out. It didn't help my grade in his class but I was pleased to give my low opinion of education departments in college. It is my opinion that math teachers should be taking far more math classes. They should know their subject and know how to implement puzzles and enrichment material in their classes.

Having taught math for so many years, I think I am in the position to make very good suggestions on improving math education in America. I have decided first to look at a weaker student.

Almost every high school math teacher can tell you about the unpleasant experience of teaching remedial or general math. By and large, it was a terrible experience for both teacher and student. I would recommend a careful review of being more effective than the expensive and unproductive remedial classes.

It is clear that the crucial years of learning elementary math happens in grades 1-4. If a student falls behind by grade four, a conference should be set up with parents informing them of the implications of this deficiency. Let's say our weak math student at age 14 has fallen through whatever safety nets grades 1-8 have provided and here he is as a freshman in high school.

Here is what I recommend: a system of rewards and punishment put in place to motivate the student in the right direction. A minimum level of competency exams should be set up that the student myust pass in order to graduate from high school. Our student should prepare for these three levels of minimum competency exams by taking numerous practice exams. In order to help students along, teachers should be given a group of students to help and motivate the weaker student. Rewards and recognition should be available so the pattern of failure is changed.

The three competency exams are as follows:

Exam I: the four operations, knowing the multiplication tables

Exam II: some understanding of fractions and percentages in money problems

Exam III: simple area, simple graphs, knowing how to use a calculator and simple familiarity with computers.

In my opinion a 14-year-old student who is four or more years behind in math is a student at risk to get into big trouble. This is the type of student who is in most need of guidance. Give this student a plan to succeed in passing Exams I, II, and III. Make it very clear what these exams are about so the student can prepare for these exams.

If the student fails to pass these three exams, not only did he fail but THE SCHOOL failed him. I am a firm believer in accountability--schools and teachers should be judged on how students do on standardized exams.

If I had a son who was a very weak student, I would make every effort to identify their talents and prepare them early in life to establish marketable skills. It is important that these non-academic students get involved in projects which give them a feeling of success, be it in helping others, community projects, even planting a garden in a local school. There are many ways a student can get satisfaction and the feeling of doing something good that aren't academic.

For average students

I was always impressed with the New York State Regents' Exams. The exams clearly outline what skills are required of the student. The exams also help the student and the teacher in organizing the subject material. Students, teachers, and the school are more motivated when they are confronted with this healthy dose of competition.

For the better student

Each school system should have a math co-ordinator to help form math clubs, math contests to help teachers in presenting puzzles and enrichment material, and to help in giving direction to the gifted student.

The National Science Foundation should have a national program to identify outstanding math and science students and provide special summer programs for our gifted students.

Excellent students need encouragement and direction. With the technical teaching tools available, excellent students can now move ahead at their own pace. In my opinion, such a program is of vital importance in upgrading math education in America.

Laszlo Ratz

I know that most of my readers have never heard of John Von Neumann (the greatest mathematician of the 20th century), Eugene Wigner (Nobel Prize-winning physicist), Edward Teller (father of the hydrogen bomb), Leo Szilard, and Paul Erdos. These men were major figures in the shaping of the 20th century. Unbelievably, all of these men went to the same small high school in Budapest: the Lutheran Gymnasium. They were all very much influenced by their math teacher Laszlo Ratz. Ratz was born in 1863 in Hungary, and he studied mathematics at the Academy of Science in Budapest. Afterwards he became a professor Evangelical High School of Budapest (also known as the Lutheran Gymnasium), where he taught mathematics until 1925. Ratz died in Budapest in 1930.

Hardly anyone has ever heard of him, but the story of Laszlo Ratz and his students would make an excellent book or movie. My first choice to be an author of such a book in my opinion would be the outstanding writer and historian Richard Rhodes, author of "Making the Atomic Bomb." I sent him a note encouraging him to undertake such a book. I am rather confident it would be a great success.

The Famous Letter to FDR

{

On July 14, 1939, Eugene Wigner and Leo Szilard drove an old Dodge (Leo Szilard never ever drove a car) to Peconic, Long Island to see Albert Einstein about a matter of grave importance. The two Nobel Prize-winning physicists got lost for several hours but eventually found Einstein. They urged Einstein to send a letter to the Belgian Ambassador regarding the possibility of the dangers of a nuclear weapon. At the time, the Belgian Congo was the only known source of uranium.

On July 31, 1939, Leo Szilard was of the opinion that the matter was of such great importance that FDR should be informed. So on July 31, 1939, Leo Szilard set out with a young physics professor from Columbia University, Edward Teller, to Einstein's home to convince him to write a letter to FDR. What followed is perhaps the most important letter of the 20th Century, from Albert Einstein to FDR:

Sir:

Some recent work by E. Fermi and L. Szilard, which has been communicated to me in manuscript, leads me to expect that the element uranium may be turned into a new and important source of energy in the immediate future. Certain aspects of the situation which has arisen seem to call for watchfulness and, if necessary, quick action on the part of the Administration. I believe therefore that it is my duty to bring to your attention the following facts and recommendations:

In the course of the last four months it has been made probable - through the work of Joliot in France as well as Fermi and Szilard in America - that it may become possible to set up a nuclear chain reaction in a large mass of uranium, by which vast amounts of power and large quantities of new radium-like elements would be generated. Now it appears almost certain that this could be achieved in the immediate future.

This new phenomenon would also lead to the construction of bombs, and it is conceivable - though much less certain - that extremely powerful bombs of a new type may thus be constructed. A single bomb of this type, carried by boat and exploded in a port, might very well destroy the whole port together with some of the surrounding territory. However, such bombs might very well prove to be too heavy for transportation by air.

The United States has only very poor ores of uranium in moderate quantities. There is some good ore in Canada and the former Czechoslovakia, while the most important source of uranium is Belgian Congo.

In view of the situation you may think it desirable to have more permanent contact maintained between the Administration and the group of physicists working on chain reactions in America. One possible way of achieving this might be for you to entrust with this task a person who has your confidence and who could perhaps serve in an inofficial capacity. His task might comprise the following:

a) to approach Government Departments, keep them informed of the further development, and put forward recommendations for Government action, giving particular attention to the problem of securing a supply of uranium ore for the United States;

b) to speed up the experimental work, which is at present being carried on within the limits of the budgets of University laboratories, by providing funds, if such funds be required, through his contacts with y private persons who are willing to make contributions for this cause, and perhaps also by obtaining the co-operation of industrial laboratories which have the necessary equipment.

I understand that Germany has actually stopped the sale of uranium from the Czechoslovakian mines which she has taken over. That she should have taken such early action might perhaps be understood on the ground that the son of the German Under-Secretary of State, von Weizsäcker, is attached to the Kaiser-Wilhelm-Institut in Berlin where some of the American work on uranium is now being repeated.

Yours very truly,

Albert Einstein

Weingartner Global Initiative

I am now 74 years old, and America has been wonderful to me and my family. I wanted to give something back to the USA, so in 2008, I funded a program at the College of William and Mary in Williamsburg, Virginia called the Weingartner Global Initiative.

There are three different outcomes for the near future:

a) a terrible catastrophe will befall the people on our planet and there is nothing we can do about it

b) nothing terrible will happen no matter what we do.

c) the ambiguous case is what this program is all about. With proper guidance, we can increase the probability of averting a catastrophe.

The plan is designed to unite students both domestically and world-wide to offer solutions to major global problems. I would like to see united students be the engine of change in improving the chances of survival of our good planet Earth.

The program has started with some very able and enthusiastic students and faculty of the College of Williams and Mary. I would like to see William and Mary be the flagship university to play a role on global issues, as it did in the 1760s.

Immediate Plans

a) starting in September 2010 with our core group of five students, we will begin to establish contact with other students from the US and other foreign universities.

b) we will try and get William and Mary to host a group of five students from both Israel and Palestine to attempt to hammer out a solution to the long-standing Israeli-Palestinian dispute. William and Mary has some excellent contacts in Washington, DC. They are presently very close to a solution and in my opinion if a solution came from some other direction than government officials, it has a good chance of success.

c) we will try and get William and Mary to host a group of young religious leaders to meet at William and Mary to try and find a way to lessen the obvious tensions between different religious groups.

I plan to use any and all profits from this book to fund this project. The following captures the spirit of the Weingartner Global Initiative.

To this end, would you join us in a peaceful, secular, non-political endeavor?

Well, this has been some chapter, but now let's do some math.

Here are 20 problems for you to try. With the help of a friend, try and get ten or more correct. Answers follow the 20 problems.

1. My aunt is very touchy about her age. She claims she is 20, not counting Saturday or Sunday. How old is she?
2. What is the smallest number divisible by 2, 3, 4, 5, and 6?
3. A man spends 1/3 of what is in his wallet, and then 1/3 of what is left. He spent $10 total. What was originally in his wallet?
4. Divide 100 marbles between Abe, Bea, and Kay. Bea gets 10 more marbles than Abe. Kay gets 20 more marbles than Bea. How many marbles does each get?
5. If I save

   $0.01 the first day

   $0.02 the second day

   $0.04 the third day

$0.08 the fourth day

$0.16 the fifth day

...and so on, how much money will I save after 10 days?

6. A cup of coffee and a piece of pie costs $3. If the pie costs $2 more than the cup of coffee, what does each cost?

7. A woman has 3 daughters, who in turn each has 3 daughters. How many pairs of sisters are there?

8. A 1/2 mile train traveling at 60 mph enters a 2 mile tunnel. How long will it take for the train to pass completely through the tunnel?

9. When I was 14, my mother was 41. She is now twice as old as I am. How old am I?

10. A man determined to stop smoking decides to only pick up cigarette ends. He can make one cigarette out of 4 ends. One day he collects 16 ends. How many cigarettes can he smoke?

11. If 1 costs $1

    12 cost $2

    144 cost $3

    What is being sold?

12. A clock rings once at 1 o'clock

    twice at 2 o'clock

    three times at 3 o'clock, etc.

What would be the total number of rings in a whole day?

13. 1% of 1% of one million equals what?

14. Mr. Baker made some pizza dough using extra strong yeast. When left in a warm place, it always doubles in size every 24 hours. If it takes 4 days to rise to the top of a very large bowl, how long would it take to be exactly halfway up?
15. What is the sum of the largest plus the smallest three-digit number you can get by rearranging the digits 105?
16. A man has a 7 gallon bucket and a 4 gallon bucket goes to a well to get exactly 5 gallons of water. How does he do it?
17. Mary and Werner each have gold coins. Mary says "If you give me one of your coins, we will have an equal number of coins." Werner says "If you give me one of your coins then I will have twice as many coins as you have." How many gold coins does each have?
18. Find two different numbers between 1 and 10 such that their product plus their sum equals 35.
19. Find the smallest number that is divisible by 7 but has a remainder when divided by 2, 3, 4, or 5.
20. You deserve a rest. Why not call up a friend and tell them about this exciting book? Everyone gets #20 right!

Answers

1. 28
2. 60
3. 18
4. 20, 30, 50
5. $10.23
6. $2.50 and $0.50

7. 12

8. 2.5 minutes

9. 27

10. 5

11. they are selling numbers

12. 156

13. 100

14. 3

15. 510 + 105 = 615

16. trial and error

17. 5 and 7

18. 8 and 3

19. 49

Before we leave Chapter 2, here are three little tests for you to enjoy. Answers after each test.

Test I

1. A couple has 5 sons. Each son has 1 sister. How many children are there?

2. If 5x=x, what is x?

3. If 5x=x and a student cancels out x and gets 5=1, what went wrong?

4. Do scientists know how gravity works?

5. Which statement is false:

    a) no president dies in the 18th century

    b) President Tyler has a living grandson

c) Aaron Burr was the third vice-president.

6. What happens once a minute, twice a week, and once a year?

7. My brother Herbert claims he can walk on water. How?

8. Solve for x: $5/0.01 = x + 1/2$

9. Find the sum of all odd numbers between:

a) 1 to 7

b) 1 to 9

c) 1 to 11

d) 1 to 13

10. Do you see a pattern in the above question?

Bonus question: Find the sum of all odd numbers from 1 to 29.

Test I Answers

1. 6

2. x=0

3. You are not allowed to divide by zero.

4. No

5. George Washington died in December 1799. I know President Tyler's grandson Harrison, and he is a nice, friendly man

6. The letter e

7. Herbert can walk on ice

8. 499.5

9. a) 16

   b) 25

   c) 36

   d) 49

10. They're all perfect squares

Bonus Question: 225

Rating

0-2: This book is not for you

3-4: still hope

5-7: average

8-9: excellent

10: you're a genius

Test II

1. A man has $1,000 in stocks. The first year, his stocks go down in value 50%. The next year the stock increases in value by 50%. What is the stock worth after 2 years?
2. If x(x-4)=0, what are the two values of x?
3. The sum of two numbers is 20. What is their maximum product?
4. What is their minimum product?
5. An English teacher and a history teacher go into a store that has a 20% discount and a 10% senior discount. The history teacher tells the clerk why not just take 30% off. The English teacher says "Obviously, I will take the 20% discount first." The clerk says the order doesn't matter. Who is right?
6. Four tennis players meet at a tournament. Each player is to play 1 game with each of the other players. How many games are there?
7. Take all the numbers 0 to 9. Which is greater, adding them or multiplying them?
8. The sum of 3 numbers equals the product of 3 numbers. What are the numbers?
9. This is a famous problem from the past. A father stipulates in a will that the oldest son gets 1/2 of the livestock, the middle son gets 1/4, and the youngest son gets 1/5. There are 19 cows and the sons couldn't find a way to divide 19 cows. A neighbor lends them one of his cows. Now the cows are divided: 1/2 of 20 = 10, 1/4 of 20 = 5, and 1/5 of 20 = 4. 10 + 5 + 4 = 19 cows, and they return the cow to their neighbor. The 3 sons were happy this problem could so easily be solved, but they couldn't explain what happened. Can you?
10. A car travels from point A to point B at 60 mph. Then he returns from point B to point A at 40 mph. What is his average speed?

Bonus Question: Uncle Werner gives each of his nephews Abe and Ben $1,000. Abe puts it into a savings account with 6% annual interest. Ben invests his money into the stock market. Ben's investments increase in value by 50% the first year, and go down in value by 40% the next year. After two years, how much does each boy have?

Answers to Test II

1. \$750

2. 0 and 4

3. 10x10=100

4. 1x19=19

5. The clerk

6. 6

7. sum of 0 to 9 is 45; their product is 0

8. 1, 2, 3

9. 1/2 + 1/4 + 1/5 = 19/20, NOT 20/20

10. say A to B is 120 miles. Trip from A to B takes 2 hours. B to A takes 3 hours. 240/5=48 mph

Bonus Question: Abe 1,000(1.06) = 1,060 after one year

1,060(1.06) = 1123.60 after two years

Ben \$1,000 up 50% = \$1,500 after one year

down 40%: 1,500 x 0.60 = \$900 after two years

Test III

1. The sum of two numbers is 20. Their difference is 6. What is the product of the two numbers?

2. Using the numbers 1, 2, and 3, how many 3 digit numbers can you write by rearranging the numbers?

3. Find the sum of the three-digit numbers you can get by rearranging the numbers of 210.

4. Decode *Nbui jt gvo.*

5. A pamphlet has 5 leaves making up 20 pages. What page numbers are on the second leaf?

6. Rearrange the letters in the word "plea" to make three new words. *Note: use all four letters in your new words!*

7. There are four different years in the 20th century in which sum of the digits of the year is 13. What are the four different years?

8. Find two different numbers between 1 and 10 such that the sum of the numbers added to the product is 34.

9. Find four four-letter words by rearranging the letters of the word "meta."

10. What comes next: o t t f f s s e ? (Hint: start counting)

Answers to Test III

1. 7x13=91

2. 6

3. 210+201+120+102=633

4. Math is fun

5. 3, 4, 17, 18

6. leap, pale, peal

7. 44x44=1936

8. 1903, 1912, 1921, 1930

9. 6 and 4

10. meat, tame, mate, team

Bonus Question: they are the first letters of numbers. 9=n

Chapter II Surprise: How I almost failed Art I in 1953

I was a freshman at the City College of New York taking Art I. The art teacher introduced the work of Jackson Pollack, and I had the nerve to express a very negative reaction to his work. The teacher asked me to sit down and said that I had no idea what I was talking about. I insisted on saying that Jackson Pollack and his sponsor Peggy Guggenheim perpetrated a giant fraud on society. Again, the teacher asked me to sit down and stated clearly that I had no idea of what I was talking about. I almost failed Art I, but 57 years later, I still abhor this type of art and I don't regret expressing my point of view.

CHAPTER III

10 MINI TESTS

Test I

Magic Square 15

Using the numbers from 1-9, fill in the boxes such that every column, row and diagonal adds up to 15.

	5	

Magic Square 34

Using the numbers from 1-16, fill in boxes such that each column, row and diagonal adds up to 34.

1			4
		10	
13			16

Test II

1) A bat and a ball cost $11, the bat cost $10 more than the ball. What does the ball cost?

2) If Washington + Lincoln = 6. Lincoln + Lincoln = _?_

3) A and C are 2 yards apart. An ant travels towards C at 2 inches per minute. A cockroach travels towards A at 6 inches per minute. How long does it take for the two bugs to meet?

4) When the ant and the cockroach do meet, which is closer to A?

5) A new home owner bought items at a store: 1 cost $1…23 cost $2…456 cost $3. Explain.

6) $\frac{4+4}{4} + \sqrt{4} = ?$

7) Find equations for 7-9, in which values for *x* and *y* are given.

x	1	2	5	9
y	3	4	7	11

x	1	2	3	4	48
y	24	12	8	6	½

x	1	3	4	5	10
y	2	10	17	26	101

What comes next in the following series: J F M A M J J A _?_

Test III

x	0	1	2	3	4	5	10
y_1	-1	4	9	14	19	24	_?_
y_2	0	1	4	9	16	25	_?_
y_3	0	1	8	27	64	125	_?_
y_4	0	2	4	8	16	32	_?_
y_5	3	4	7	12	19	28	_?_
y_6	0	2	8	18	32	50	_?_
y_7	0	16	64	144	256	400	_?_
y_8	---	48	24	16	12	9.6	_?_
y_9	---	48	12	$\frac{48}{9}$	$\frac{48}{16}$	$\frac{48}{25}$	_?_
y_{10}	0	1	10	11	100	101	_?_

Test IV

1– 9.) Find equations of y_1 through to y_9

10.) Explain the equation for y_{10}.

Test V

1.) Find a 3 digit number such that if you turn it upside down it is a source of energy.

2.) A man having a 7 gallon measure and a 4 gallon measure goes to a well and gets exactly 5 gallons of water. How?

3.) Ted meets his wife every day at the train station at 5pm. She starts at 4pm and they get home at 6pm. One day he gets to the station early and starts walking. He meets his wife on the road and they get home at 5.50pm. How much time did he walk?

4.) A man sells 2 horses each for $99. If he gains 10% on one and loses 10% on the other, what are his profits?

5.) A party of 8 order a large pizza. The waiter cuts the pie 3 times to get 8 slices. How?

6, 7.) This counts for 2 questions. Very few people get it. Hint: it has something to do with arithmetic.

O T T F F S S E _?_

8.) When Frank was 16, he was 6 feet tall and a tree was marked at 6 feet. The tree grows 1 inch per year. How high is the mark when Frank is 36 years old?

9, 10.) This counts for 2 questions. Assume the earth is a perfect sphere. A band is placed around the equator. If 3 feet were added to the band, how far off the ground would it be? Assume the radius of the earth is 4,000 miles.

Test VI

1.) (4 +.4+square root 4)/.4

2.) 4!=4 X 3 X 2 X 1=24

Find 4!/.4

3.) In a school of less than 200 students, one day is set aside each year as Fun Day.

$\frac{1}{2}$ of the students go to the zoo

$\frac{2}{3}$ of the rest go to an Irving Berlin Concert

$\frac{3}{4}$ of the rest go to an Elvis show

$\frac{4}{5}$ of the rest go to the Chess Club

One student is left and he goes to the Math Club. How many students participated?

4.) What is the probability that the students ever heard of Irving Berlin, Richard Rodgers or Oscar Hammerstein?

5.) Prove $\sqrt[3]{2} > \sqrt[10]{10}$

6.) The angles of a triangle are in the ratio of 1:2:3. Find the ratio of sides.

7.) Easy question: the ratio of the angles of a triangle is in the ratio of 1:1:2. Find the ratio of sides.

8.) I hope you are enjoying this book. The year I was born is a perfect square. What year was I born?

9.) 2 whole numbers add up to 20.

Find the maximum product

Find the minimum product

10.) In a barn there are people and cows. There are 22 heads and 72 feet. How many cows are there?

Test VII

1.) Which number when added to $1\frac{1}{4}$ equals the same number multiplied by $1\frac{1}{4}$?

2.) Two US coins have a combined value of 55¢. One coin is not a nickel. What are the two coins?

3.) A very important mathematical law was used incorrectly. Find the error.

Given: $a = b$

$ab = b^2$

$a^2 - ab = a^2 - b^2$

$a(a - b) = (a + b)(a - b)$

$a = a + b$

$a = 2a$

$a = 1$

4.) Find the smallest number divisible by 1, 2, 3, 4, 5, 6, 7, 8, 9, 10

5.) Find th4e longest pole that will fit in a 2” cube.

6.)How many diagonals are there in a hexagon?

7.) $2^{10}=10^3+ X^2-1$ X=?

8.,9.,10.) This is a rather important law to know . The Rule of 72. To find the length of time it takes money to double in value, divide 72 by the rate of interest. For example at 6% interest, it will take 12 years for money to double. At 8%, the money will double in 9 years.

In 1830 a relative puts $1 in the bank at 8% interest. How much money is in the bank in 2010?

Test VIII

1.) This test has only one hard question which I would like you to try on your own. This problem will be discussed in Chapter V. Diophantine equations.

A woman wants to buy 100 presents for $100.00 Pens are 10 for $1.00, Irving Berlin CD's are $2.00 each, this book is $5.00 each. How many of each does she buy?

Test IX

The binary system has taken on a very important rokle in understanding the so called digital age.

First let us look at our base 10.

$749 = 7(10^2) + 4(10) + 9$

$8063 = 8(10^3) + 0(10^2) + 6(10) + 3$

Now in bas 2 we have only a 0 or1. Let us count. Remember you can use only either 1 Or 0.

0,1,10,11,100,101,111,1000,1001,etc.

Note our 2=10 base 2

4=100 base 2

8=1000 base 2

16=10,000 base 2

Our number $13 = 1(2^3) + 1(2^2) + 0(2) + 1 = 1101$

Going back the other way, from base 2 to base 10

$1101 = 1(8) + 1(4) + 0(2) + 1$

$19 + 1(2^4) + 0(2^3) + 0(2^2) + 1(2) + 1 = 10{,}011$

Going the other way

$10{,}011 = 1(16) + 0(8) + 0(4) + 1(2) + 1 = 19$

Now convert the following numbers to base 2 and then back to base 10.

5,12,21,100

Now lets make up an addition and multiplication table for base 2.

0+0=0	0+1=1	1+0=1	1+1=10
0 X 0=0	0 X 1=0	1 X 0=0	1 X 1=1

There are many applications in base 2, it is very important. For example

0=No 1=YES

0=light is off 1=light is on

0=no current 1=current

Every number in base 10 can be converted to base 2.

Let L=light is on

N= no light

LLNL=1101

LNN=100

9b.) Find 5 fractions, more than a ¼ but less than a ½.

9c.) Prove there are an infinite number of fractions between ¼ and ½.

¼=.2500….

½=.5000…

.26,.27,.28, etc.

Test X

You need a rest so why not go over these questions with a friend. Get ready for more fun in Chapter IV with more math and other memorable stories.

Are you having fun?

ANSWERS FOR TEST 1

2	9	4
7	5	3
6	1	8

There are many solutions

1	14	15	4
8	11	10	5
12	7	6	9
13	2	3	16

ANSWERS FOR TEST 2

Bat $10.50, Ball $.50

5+5=10

9 minutes

Same distance from A

Buying numbers

13

$Y=x+2$

$Y=\frac{24}{x}$

$Y=x^2+1$

Months S=September

ANSWERS FOR TEST 3 and 4

$Y=5x-1$	$X=10$	$Y=49$
$Y=x^2$	$X=10$	$Y=100$
$Y=x^3$	$X=10$	$Y=1000$
$Y=2^x$	$X=10$	$Y=1024$

There is a mistake when x=0 and y=1, did you spot the error?

$Y=x^2+3$	$X=10$	$Y=103$
$Y=2x^2$	$X=10$	$Y=200$
$Y=16x^2$	$X=10$	$Y=1600$
$Y=\frac{\text{[illegible]}}{x}$	$X=10$	$Y=\frac{\text{[illegible]}}{10}=.48$
$Y=\frac{48}{x^2}$	$X=10$	$Y=\frac{48}{100}=.48$

Base 2 10=1010 in base 2

ANSWERS FOR TEST 5

1. OIL= 710

2. Easy

3. She saved 10 minutes, which is 5 minutes each way. (55 minutes)

4. He paid \$90 & \$110 for the horses. (Lost \$2)

5. He places one piece on top.

6. Count one=o, two =t, nine=n

7. See above

8. The tree grows but the mark is always 6’

9.,10. The first band= $2\pi R=8000\pi$ if we add 3’. The new length is $8000\pi+3 = 2\pi R$. The new radius is $\frac{8000\pi+3}{2\pi}=4000+\frac{3}{2\pi}=R$. So it is $\frac{3}{2\pi}$ off the ground=about $\frac{1}{2}$

ANSWERS FOR TEST 6

1. 16
2. 60
3. 120 students
4. Very low-Irving Berlin, Richard Rogers and Oscar Hammerstein deserve more attention from young people.
5. Raise each 30 power. 1024>1000
6. 30-60-90 Δ Sides are 1:√3:2
7. 45-45-90 Δ Sides are 1:1:√2

8. 1936 = 44x44

9. 10x10=100 1x19=19

10. 14 cows, 8 people

ANSWERS FOR TEST 7

1. 5

2. One is not a nickel, but the other is. The two coins are a 50¢ piece and a 5¢ piece.

3. You are not allowed to divide by zero.

4. 8x9x7x5=2520

5. Square root of 12

6. 15.

7. x=5

8. 9. 10. 2010-1830=180years 72/8=money doubles every 9 years So 180/9=20 So money will double 20 times. The answer is 1,048,576.

END OF CHAPTER 3 SURPRISE

My 20 favorite movies in order of preference:

1) Gone With the Wind
2) Wizard of Oz
3) Singing in the Rain
4) All Quiet on the Western Front
5) Casablanca
6) Crimes and Misdemeanors
7) Hannah and Her Sisters

8) Young Frankenstein
9) Forrest Gump
10) Some Like it Hot
11) Godfather series
12) High Noon
13) Snow White and the Seven Dwarves
14) Blazing Saddles
15) On Golden Pond
16) Driving Miss Daisy
17) Blue Angel
18) Marty
19) An American in Paris
20) Das Boat

PS- my favorite movie is the movie I made, "1+1=X." Unfortunately, the rest of the world did not share my opinion of this movie. If you would like to see a 9 minute trailer of the movie, search for "Werner Weingartner" on the Internet.

CHAPTER IV

THREE VERY INTERESTING PROBLEMS YOU SHOULD SHARE WITH YOUR FRIENDS.

Problem #1: President James Garfield 20th original proof of the Pythagorean Theorem

Several Presidents stand out as excellent mathematicians. They include President James Garfield, President U.S. Grant and President George Washington. Without a doubt President Garfield is on top of this list. At 16, he was a laborer on the canals and at age 23 he was President of a college. He was able to write Latin with his right hand and at the same time write Greek with his left. Since this is mainly a math book I would like to present President Garfield's original proof of the Pythagorean Theorem, which in my opinion is a real gem:

This proof, discovered by President J.A. Garfield in 1876 is a variation on the previous one. But this time we draw no squares at all. The key now is the formula for the area of a trapezoid - *half sum of the bases times the altitude* - (**a** + **b**)/2·(**a** + **b**). Looking at the picture another way, this also can be computed as the sum of areas of the three triangles - **ab**/2 + **ab**/2 + **c**·**c**/2. As before, simplifications yield $\mathbf{a^2 + b^2 = c^2}$.

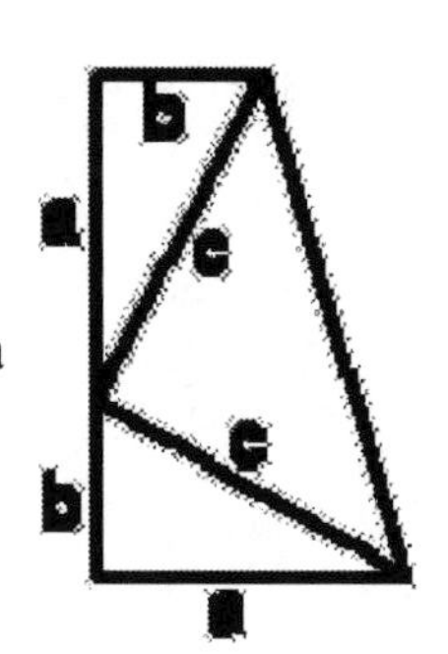

Area of Trapezoid= $\frac{1}{2}h(base_1 + base_2)$ $\frac{1}{2}(a+b)(a+b) = \frac{1}{2}ab + \frac{1}{2}ab + \frac{1}{2}c^2$

$$(a+b)(a+b) = ab + ab + c^2$$

$$a^2 + 2ab + b^2 = 2ab + c^2$$

$$x^2 + y^2 = z^2$$

Problem #2: The 4-4 Problem

This problem was first introduced to me by Irving Yano on a much simpler level. It is one of my favorite problems but very difficult. Using four 4s, write the numbers from 1-100 using all of four 4s. You are allowed to use +, -, ×, ÷, .4, 44, $\sqrt{4}$, $.\overline{4}$ (= .444… = $\frac{4}{9}$) so that $\sqrt{.\overline{4}} = \frac{2}{3}$

4! = 4×3×2×1 = 24 $.\sqrt{4} = .2$

I have never seen any student or teacher get all 100. In 4 hours try and get about 60 out of 100. 70 out of 100 is very good; 71- 90 is excellent and 95 or more WOW!

For example:

$7 = 4 + \frac{4}{4} + \sqrt{4}$ $17 = \frac{4 + \sqrt{4}}{.4} + \sqrt{4}$

$17 = 4 \times 4 + \frac{4}{4}$ $37 = \frac{4!}{\sqrt{.\overline{4}}} + \frac{4}{4}$

$45 = \frac{4! - 4 - \sqrt{4}}{.4}$ or $\left(\frac{\sqrt{4}}{.4}\right)\left(\frac{4}{.\overline{4}}\right)$ or $44 + \frac{4}{4}$

$61 = \frac{4!}{.4} + \frac{4}{4}$ $87 = 4 \times 4! - \frac{4}{.\overline{4}}$

$81 = \frac{4 \times 4 + .\sqrt{4}}{.\sqrt{4}}$ or $\left(\frac{4 - .4}{.4}\right)^{\sqrt{4}}$

89 is very difficult… $89 = \frac{\frac{4!}{\sqrt{.\overline{4}}} - .4}{.4}$

As you can see you must be patient ingenious and know your numbers to do this problem.

CHAPTER IV

THREE VERY INTERESTING PROBLEMS YOU SHOULD SHARE WITH YOUR FRIENDS.

Problem #1: President James Garfield 20th original proof of the Pythagorean Theorem

Several Presidents stand out as excellent mathematicians. They include President James Garfield, President U.S. Grant and President George Washington. Without a doubt President Garfield is on top of this list. At 16, he was a laborer on the canals and at age 23 he was President of a college. He was able to write Latin with his right hand and at the same time write Greek with his left. Since this is mainly a math book I would like to present President Garfield's original proof of the Pythagorean Theorem, which in my opinion is a real gem:

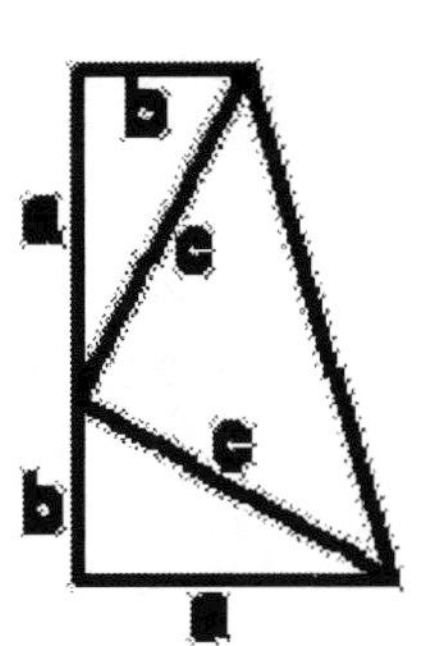

This proof, discovered by President J.A. Garfield in 1876 is a variation on the previous one. But this time we draw no squares at all. The key now is the formula for the area of a trapezoid - *half sum of the bases times the altitude* - (**a** + **b**)/2·(**a** + **b**). Looking at the picture another way, this also can be computed as the sum of areas of the three triangles - **ab**/2 + **ab**/2 + **c**·**c**/2. As before, simplifications yield $\mathbf{a^2 + b^2 = c^2}$.

Area of Trapezoid= $\frac{1}{2}h(base_1 + base_2)$ $\frac{1}{2}(a+b)(a+b) = \frac{1}{2}ab + \frac{1}{2}ab + \frac{1}{2}c^2$

$$(a+b)(a+b) = ab + ab + c^2$$

$$a^2 + 2ab + b^2 = 2ab + c^2$$

$$a^2 + b^2 = c^2$$

Problem #2: The 4-4 Problem

This problem was first introduced to me by Irving Yano on a much simpler level. It is one of my favorite problems but very difficult. Using four 4s, write the numbers from 1-100 using all of four 4s. You are allowed to use +, -, ×, ÷, .4, 44, $\sqrt{4}$, $.\overline{4}$ (= .444... = $\frac{4}{9}$) so that $\sqrt{.\overline{4}} = \frac{2}{3}$

4! = 4×3×2×1 = 24 $\quad .\sqrt{4} = .2$

I have never seen any student or teacher get all 100. In 4 hours try and get about 60 out of 100. 70 out of 100 is very good; 71- 90 is excellent and 95 or more WOW!

For example:

$7 = 4 + \frac{4}{4} + \sqrt{4}$ $\qquad 17 = \frac{4 + \sqrt{4}}{.4} + \sqrt{4}$

$17 = 4 \times 4 + \frac{4}{4}$ $\qquad 37 = \frac{4!}{\sqrt{.\overline{4}}} + \frac{4}{4}$

$45 = \frac{4! - 4 - \sqrt{4}}{.4}$ or $\left(\frac{\sqrt{4}}{.4}\right)\left(\frac{4}{.\overline{4}}\right)$ or $44 + \frac{4}{4}$

$61 = \frac{4!}{.4} + \frac{4}{4}$ $\qquad 87 = 4 \times 4! - \frac{4}{.\overline{4}}$

$81 = \frac{4 \times 4 + .\sqrt{4}}{.\sqrt{4}}$ or $\left(\frac{4 - .4}{.4}\right)^{\sqrt{4}}$

89 is very difficult… $89 = \frac{\frac{4!}{\sqrt{.\overline{4}}} - .4}{.4}$

As you can see you must be patient ingenious and know your numbers to do this problem.

Problem #3: The Euler Line

One of the most beautiful and pleasing problems, in my opinion, in mathematics is the Euler line. It combines algebra, geometry and has a truly aesthetic touch to it that is memorable. Well you be the judge.

Let me illustrate with the following example:

Given any triangle

The altitudes meet at one point A

The 3 medians meet at one point B

The 3 perpendicular bisectors meet at one point C

A, B and C are always on a straight line called the Euler line.

Let me illustrate with a problem. Before I start, I hope the readers know if 2 lines are perpendicular then there slopes are negative reciprocals. For example, if a line has a slope of $\frac{2}{3}$, then the perpendicular line has a slope of $-\frac{3}{2}$

Now consider the triangle RST

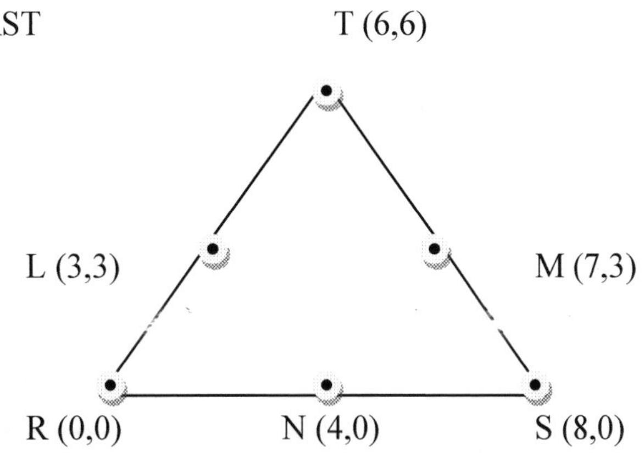

T (6,6)

L (3,3) M (7,3)

R (0,0) N (4,0) S (8,0)

The mid points re L (3,3); M (7,3); N (4,0).

Step 1—Where do the altitudes meet?

From T the altitude is X= 6

From R, note the slope of TS = -3 so from R the altitude has a slope of $\frac{1}{3}$. The equation altitude is $y = \frac{1}{3}x$

From S note the slope of RT = 1 so altitude from S $y = -1x + b$ (8,0)

$0 = -1(8) + b$ $\quad b = 8$ $\quad y = -1x + 8$

It is easy to show all 3 lines meet at (6,2) point A.

Step 2—Where the 3 medians meet:

median RM equation is $y = \frac{3}{7}x$

median TN (6,6) (4,0) $\quad y = 3x + b$ use (6,6) $\quad 6 = 18 + b \quad b = -12$

$y = 3x – 12$

median SL (8,0) (3,3) $\quad$ slope $-\frac{3}{5}$ $\quad y = -\frac{3}{5}x + b$ use (8,0)

$0 = -\frac{3}{5}(8) + b$ $\quad b = \frac{24}{5}$ $\quad y = -\frac{3}{5}x + \frac{24}{5}$

$y = \frac{3}{7}x$ $\quad$ and $y = 3x - 12$ $\quad \frac{3}{7}x = 3x - 12$

$x = 4\frac{2}{3}$ $\quad y = 2$

They all meet at $(4\frac{2}{3}, 2)$ B

Step 3—3 perpendicular bisector

Perpendicular bisector at N is x = 4

Perpendicular bisector at L $y = -1x + b$ (3,3)

$3 = -3 + b$ $b = 6$ $y = -1x + 6$

Perpendicular bisector at M The slope of TS is -3 so perpendicular bis at M has a slope of $+\frac{1}{3}$

$y = \frac{1}{3}x + b$ (7,3) $3 = \frac{7}{3} + b$ $y = -\frac{1}{3}x + \frac{2}{3}$

so it is easy to show all 3 perpendicular bis meet at (4,2)C

Step 4—Are A, B and C on the same straight line.

A = (6,2) B = $(4\frac{2}{3},2)$ C = (4,2)

The Euler for this problem is $y = 2$

Now you try this problem.

Take the triangle RST T (4,4)

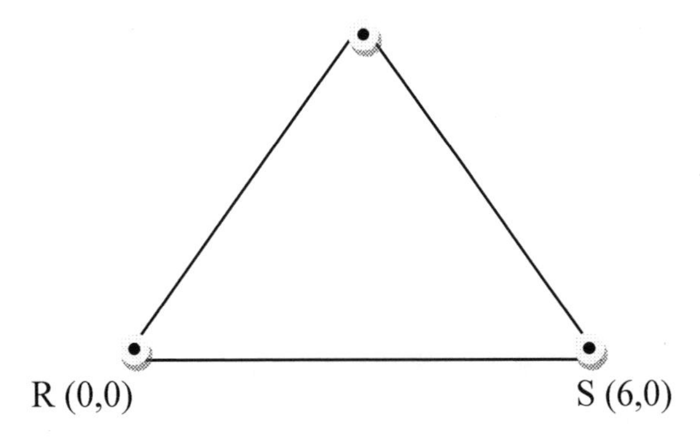

Prove 3 altitudes meet at one point A

Prove 3 medians meet at one point B

3 perpendicular bisector meet at one point C

Prove points A, B and C are on the same straight line and write the equation of the Euler line.

A much harder problem is to prove the general theorem—here we use letters to write the coordinates of R, S,T

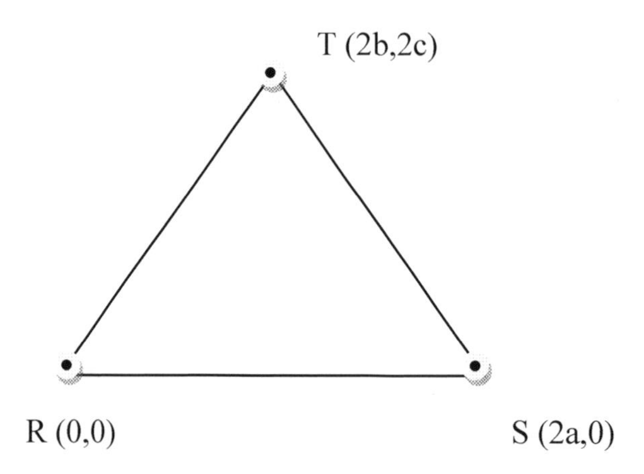

Keep in mind RT has a slope $= \frac{2c}{2b} = \frac{c}{b}$ and a line perpendicular to RT has a slope of $-\frac{b}{c}$

This is a long, hard problem. You have to be very steady in algebra but the answers are:

Point A $\left(2b, \frac{2ba - 2b^2}{c}\right)$

Point B $\left(\frac{2b + 2a}{3}, \frac{2c}{3}\right)$

Point C $\left(a, \frac{b^2 - ba + c^2}{c}\right)$

It is ever so easy to make a mistake in these calculations.

4-4 Problem:

The answers to this problem can be found online.

Euler line problem:

Triangle RST

3 altitudes meet at (4,2)

3 medians meet (10/3,4/3)

3 perpendicular bisectors meet (3,1)

Euler line y=x-2

Probability

For the rest of the chapter, I am going to concentrate on 3 Problems in Probability.

1) The math analysis of Joe DiMaggio's 56 game hitting streak.
2) The famous birthday problem.
3) An unbelievable problem.

Problem #1:

Joe has a BA= .333.

What are the odds Joe gets a hit? 1/3

What are the odds he doesn't get a hit? 2/3

1/3+2/3=1 (Certainty)

Now consider 2 at bats. Let H=Hit and N=No Hit in 2 at bats the following can happen.

H H	$\frac{1}{3}\times\frac{1}{3}=\frac{1}{9}$
H N	$\frac{1}{3}\times\frac{2}{3}=\frac{2}{9}$
N H	$\frac{2}{3}\times\frac{1}{3}=\frac{2}{9}$

NN	$\frac{2}{3} \times \frac{2}{3} = \frac{4}{9}$
	1

What are the odds of getting 2 hits = 1/9

What are the odds of getting exactly 1 hit = 4/9

What are the odds of no hits = 4/9

Now comes a key point. What are the odds that DiMaggio gets at least one hit in two times at bat? He must avoid NN= 1- 4/9 = 5/9.

We are now ready for the Joe DiMaggio 56-game hitting streak problem. Here are the assumptions:

1) in each game, he has four times at bat

2) he is consistently batting .333

Let H=gets a hit, N=no hit. What are the odds he gets at least one hit in a game with 4 times at bat?

DiMaggio must avoid NNNN. Odds are: 2/3 x 2/3 x 2/3x 2/3 = 16/81, which equals about 1/5. So the odds of his getting at least one hit with four times at bat is 1 - 1/5 = 4/5. So in any given game, the odds of him getting at least one hit is 4/5.

Now do this 56 games in a row. The odds are $(4/5)^{56}$ = about 1/250,000.

Another very famous player of 1941 was Ted Williams, who had a .400 batting average. What are the odds of his getting at least one hit in a game?

Let H=hit=2/5, N=no hit=3/5. In order for Ted to get at least one hit in a game, he must avoid NNNN, so the odds are 3/5 x 3/5 x 3/5 x 3/5 = 81/625 = 13%

So the odds he gets at least one hit is 87%. What are the odds he gets at least one hit in

a) 2 games $0.87 \times 0.87 = 0.75$

b) 4 games $(0.87)^4 = 0.57$

c) 8 games $(0.87)^8 = 0.32$

d) 16 games $(0.87)^{16}=0.10$

e) 32 games $(0.87)^{32}=0.01$

Do you see how a little simple math can give you insight into what appears to be complicated problems?

About our assumptions: in reality, the odds are less than they appear because a) Ted Williams will rarely get four times at bat in a game because the opposing team will walk him and b) Williams will not always have a 0.400 batting average--some days he or the pitcher will be better or worse than usual.

Problem #2: The Birthday Problem

What are the odds of two people sharing a birthday on the same day?

I assure you this problem is a big hit in any group. I have tested this problem in many of my math classes and it was always a big hit. Here are some of the odds:

Case 1	23 people = 50%
Case 2	30 people = 68%
Case 3	40 people = 88%
Case 4	50 people = 96%
Case 5	60 people = almost 100%

Other birthday problems

If you have 88 people, the odds of three people sharing a birthday is 50%. In a gathering of 1,000 people, the odds that 9 of them share a birthday are better than 50%. It's hard to believe, but it's mathematically sound.

The birthday problems are examples of what is called occupancy problems, in which mathematicians think about placing balls in 365 cells. This innocent-looking birthday problem has huge applications in math and physics.

Problem #3: An Unbelievable Problem

I was in a physics class and the teacher was talking about uncertainty theory. He pointed out things frequently different from what seems possible. He gave the class the following problem which I will never forget. Let's say 1% of the population has disease C. There is a test for C that is 99% accurate. A woman goes for the test and it comes out positive. He asked what the odds are that she has disease C. When he came out with the answer ½, I couldn't believe it. I asked the teacher to repeat the problem and I couldn't believe the answer. I spent the next ½ hour working on the problem and sure enough, he was right. Here is what I did.

Take 10,000 people. Of those 10,000 people, 1% have C and 99% do not. 1% of 10,000=100 have C and 99% of 10,000 = 9,900 do not have C.

Everyone takes the test for C. Of the 100 with C who take the test, 99% of 100 = 99 will be positive. Now 9,900 don't have C. 1% 9,900 = 99 will be positive. So from 10,000 people, 99 who do have C will come out positive and 99 who do not have C will come out positive. Things are not always as they seem.

Math Projects for Students

There are 100s of topics in math books for students to consider but I would like to offer just 10.

1. Write a report on Laszlo Ratz and his students

2. The works of Leonhard Euler. I am a big Euler fan and I would love to see at least one high school form an Euler fan club

3. 1642- Monumental Year in Science- Galileo dies and Sir Isaac Newton is born.

4. Alan Turing, 1912-1954. Churchill once remarked, "In history, rarely do so many owe so much to so few." Let me say that rarely in the course of English history is so much owed to Alan Turing and he is given so little, as he was persecuted for his homosexuality. This heroic English figure committed suicide in 1954.

5. Improving the quality of math education from the student's point of view

6. Women in mathematics- consideration of the volatile topic. Do men have a biological advantage over women in math and science? I will take the big risk and offer my opinion, which is yes, there is a big difference.

7. Permutations and combination and their application in genetics, organic chemistry, game theory, and code breaking.

Now comes three real big topics...

8. Mathematics and insanity. When I was at the City College of New York in the 1950s, I knew of two outstanding math professors who in my opinion were crazy. Several movies and plays have been written on this subject.

Movies:

- *Good Will Hunting*- about a troubled genius

- *A Beautiful Mind*- about Nobel Prize-winner John Nash

Plays:

- *Proof*- about mental disorder in math (recently made into a full-length movie)
- *Pi*- a crazy number-hunter

Here is a list of brilliant mathmaticians with mental problems:

1) Ludwig Boltzman, 1844-1906
2) Georg Cantor, 1845-1918
3) Abraham DeMoivre, 1667-1754
4) Paul Erdes, 1913-1996
5) Evariste Galois, 1811-1832. Died in a duel over a prostitute.
6) Alex Grothendieck, 1928-
7) Ted Kaczynski (the Unabomber), 1942-
8) John Nash, 1928-
9) Marius Lie, 1842-1899.
10) Andre Bloch, 1893-1948. On November 17, 1917, he murdered his brother, uncle, and aunt. He was confined to a mental hospital. A model patient, he refused to go out saying "mathematics is enough for me."
11) Some of my math prof. in 1954.
12) Sir Isaac Newton, 1642-1727.
13) Pythagoras, 569-475 BC. Head of a strange cult called the Pythagoras Brotherhood and had a student drowned because he revealed that the square root of two is irrational.
14) Srinivasa Romannjan, 1887-1920. Self-taught mathematician with little or no education.

By the way, I am just an ordinary mathematician and I don't think I am crazy.

9. Are computers the best mathematicians? It is not rare that in times students out-do their teachers. Have we reached this point in mathematics? Computers have the big advantage to constantly absorb mathematical concepts 24 hours a day. When it comes to evaluating known information, computers win with ease. In many branches of math, computers are coming up with solutions and proofs way ahead of man. Only the top five chess players in the world can compete with the computer, and the computer is always improving.

There is one area of math where man can compete and that is with coming up with new ideas in math and science. But once new ideas are accepted, the computer absorbs the information and includes it in its database of knowledge. Man is put into the impossible position of having to run faster and faster to keep up with the computer.

Conclusion: Computers--the machines men have designed--will be the best mathematicians, far surpassing their creators.

10. Decision-making, man vs. computers. It should be clear to us that computers are making more and more decisions in our increasingly complicated world. In 2008, there was a major financial crisis in which I believe no one had a total overview of what was going on. Even at the risk of losing some degree of efficiency, I think man must have an overview of what is going on and have the ability override a decision made by computers.

Recently, I went to a lecture about the Cuban Missile Crisis of 1962. A very nice, quiet man at our table informed us that he was very much involved with the crisis. He told the following incredible story. He was an airborne bomber on highest alert during the crisis, armed with hydrogen bomb. If the bombers were given a certain code, they were to deliver their load of bombs over Russia. I asked him if the code could be overridden by a person. He said no. Once the code was given, there was no turning back. I am sure in the last 50 years we have come very close several times to a world catastrophe, either by error or miscalculation by Russia or America. I am sure every major government has a white paper on the probability of a huge man-made catastrophe befalling us. I don't think the general public is aware of such a paper or is the general public aware of how we have come very close to a worldwide catastrophe on several occasions.

In the movie "War Games," we see how decisions of life and death are dependent on computer decisions. Even with infallible computers, things could go wrong. Yes, we are very, very dependent on computers, but I do hope man has an overview of world events and has the ability to override a computer decision.

I would like to add a part to this section dealing with a math paper I wrote 40 years ago titled, "The most important issue of our time as a function of s." It deals with the probability of some worldwide catastrophe.

Let p_1=the probability of a group setting off a world-altering nuclear weapon is small, say 1%. n_1=99% of this not happening.

Let p_2=the probability of a world-altering chemical or biological weapon is small, say 1%. n_2=99% probability of this not happening.

Let p_3=the risk of causing an environmental catastrophe is small, say 2%. n_3=98% that it won't happen.

Each individual probability is small, but the probability of avoiding all of them is not so small. Let P=probability of avoiding all of them, which can be written as

$$P=1-n_1-n_2-n_3-...$$

The problem is very much like asking 100 respected math professors to solve 8x7. The chances of each professor solving the problem is very, very high, but the probability of all 100 getting the correct answer of 56 is not very high. There are many possible reasons why someone would not get the right answer.

It is this section of the book that I am more uncomfortable and frustrated about. What to do about this problem is a huge question mark. Do we as a people have the skill or the discipline to minimize these risk is very much in doubt. Every thinking person has thought about these issues but most people are so overwhelmed by the scope of the problem and they come to the east conclusion that what will be, will be.

It is my opinion that we as a people will have to make many changes and sacrifices (which include living much simpler lives and consuming a fraction of the resources we consume now) to give future generations the chance at life, liberty, and the pursuit of happiness.

END OF CHAPTER IV SURPRISE

Music and math have a close connection. Here are ten of my favorite musicians.

1) Puccini
2) Verdi
3) Tchaikovsky
4) Chopin
5) Beethoven
6) Irving Berlin
7) The Beatles
8) Rogers and Hammerstein
9) Andrew Lloyd Weber
10) Richard Wagner

CHAPTER V

APRIL CODY

As mentioned, I am a retired math teacher, and when the local high school asked me to mentor members of the math club, I was eager to offer my services. April Cody was president of the math club and lived a half-mile from me. we were to meet once a week, and many of my experiences formed the foundation of this book.

At our first meeting, we introduced each other and the subject of music came up. she had never heard of Irving Berlin and I had never heard of Lady Gaga--quite a generation gap. I remember that in September 1989 I mentioned to my Advanced Placement calculus class that Irving Berlin just died at the ripe old age of 101. Out of 37 students, only five students had ever heard of him. After he died, there was a patriotic concert played each July 4th at Woodlawn Cemetery in the Bronx where Berlin was buried. I always arrived about an hour before the concert to go over to Irving Berlin's grave and sing "God Bless America," as I just loved that song. I mentioned other of my favorite musicians to the math club, such as Puccini, Verdi, Rogers and Hammerstein, and got the same response--the students had never heard of them. Yes, I did hear of the Beatles (great musicians in any era), the Rolling Stones, Sting, and The Doors, but by and large, 60 years' age difference made me an outsider in much of their world. They were very friendly and polite indicating they would give Irving Berlin a chance to be part of their music world.

At our second meeting, April came up with a gem. she observed that in right triangles where the hypotenuse is one more than a leg, then the sum of that leg and the hypotenuse is always the square of the second leg, such as triangles with measurements (3, 4, 5), (5, 12, 13), and (7, 24, 25). She asked if this were always true. I said let's call it the April Conjecture, and if we can prove it, we will call it the April Theorem. Her immediate answers was that she didn't have a clue on how to prove it. Let's call the legs y and x. This makes the hypotenuse x+1, giving the triangles sides of y, x, and x+1. To prove the April Theorem, we need to show that $y^2=x+x+1=2x+1$.

Now use the Pythagorean Theorem:

$y^2+x^2=(x+1)^2$

$y^2+x^2=x^2+2x+1$

$y^2=2x+1$

So April's conjecture was right and it is now a theorem. It was a good vehicle to show how mathematicians go from an idea to a conjecture and then prove or disprove it. April asked me if there were many such triangles where $y^2=2x+1$, and with my help, we came up with several different examples.

y=3 x=4 (3, 4, 5)

y=5 x=12 (5, 12, 13)

y=7 x=24 (7, 24, 25)

y=9 x=40 (9, 40, 41)

y=11 x=60 (11, 60, 61)

There are actually an infinite number of such triangles. I pointed out to April another famous conjecture: **Fermat's Last Theorem**.

The most famous conjecture in mathematics is called Fermat's Last Theorem. In 1665, a few weeks before he died, the French mathematician Pierre de Fermat claimed that the equation $x^n+y^n=z^n$ has no whole number solutions for an n greater than 2. Clearly $x^2+y^2=z^2$ has many whole number solutions, but for over 300 years, this was the star problem in mathematics. To add spice to the story, Fermat claimed he had a wonderful proof but the margin was too small to write it out. For over 300 years, every mathematician gave attention to this problem. Here are some of the highlights:

1665- Fermat proposes the problem

1753- Euler proves it true for n=3

1908- Wolfskel offers a huge prize for solving the problem

1977- Computers were used to prove the theorem true for n less than 125,000

1993- Computers were used to prove the theorem true for n less than 4 million (notice that computers are proving to be the best mathematicians).

Then it happened:

1994- Andrew Wiles, a math professor at Princeton University, proves Fermat's Last Theorem after 7 years of solitude.

In 1982, I made the following bet with my good friend and famous math professor, Jerry Wilson. He said that in our lifetime, there will be no solution to Fermat's Last Theorem. I pointed out the obvious: that he had no chance of winning the bet, for if he were right, then he would not be alive to collect on the bet. Not every genius has common sense. Our friendship was based on our common interest in bridge and chess, and he was president of our music club. He had tremendous powers of concentration where he would spend 18 straight hours on his specialty, the soap bubble problem in the field of topology, and forget to eat.

The most famous absent-minded story involved the world-famous math and physics professor Norbert Wiener of MIT. Norbert Wiener was to attend at two-day conference at Yale. He drove to New Haven, parks his car, and attends the lectures. At the end of the conference, he forgets he drove down and takes a bus to his home in Cambridge, Massachusetts. The next day, he opens his garage and there is no car. He calls the local police to report a stolen car. The conclusion? Just because you are a genius doesn't rule out the possibility of doing incredibly stupid things.

Now for the Werner absent-minded story. In the morning, I used to go swimming in a pool about 50 yards from our house. I always took a bar of soap with me, and since I had no pockets, I put the soap between my hat and my head. One day, I was to play bridge after my swim. I rushed over to the Bridge Club meeting and sat down, when one of the women mentioned it was rude to wear a hat indoors. She removed my hat and there was this bar of soap sitting on the top of my head. What followed was an outburst of uncontrollable laughter and jokes about the absent-minded Werner.

I am now going to tell one more story that has absolutely no connection with this chapter. The famous John con Neumann was asked to explain an auto accident he had. This is what he came up with: "I was driving along on a beautiful country road, and the trees were going 60 mph when suddenly one stepped in my path."

April just loved these stories and since I have stories on every subject, we are a perfect match.

April proposed we make up a Math All-Star Team. Here is what we came up with:

1) Newton
2) Gauss
3) Archimedes
4) Euler
5) Pythagoras
6) Euclid

7) Galileo

8) Leibniz

9) John von Neumann

10) Albert Einstein

Please note this is our list and there will never be an agreement on who should be on the list. I would also like to mention there were several of our presidents who were very talented in math, such as James Garfield and Ulysses Grant. I am also going to include George Washington because my wife is a big GWU fan.

Important Ideas in Math: Irrational Numbers

A group headed by Pythagoras called the Pythagorean Brotherhood (a very bizarre group that your high school math teacher probably won't tell you about) was very much in love with numbers and it even had religious overtones. When they proved that $\sqrt{2}$ could not be expressed as a rational number (a rational number can be expressed as a/b where a and b are whole numbers), it caused a crisis in math. Pythagoras made a ruling that divulging this proof would result in the death penalty.

Prove that $\sqrt{2}$ can't be expressed as a fraction a/b where a and b are whole numbers.

Indirect proof: assume $\sqrt{2}=a/b$ where a and b are reduced to lowest terms. Now square each side

$2=a^2/b^2$

$2b^2=a^2$

This implies that a is divisible by 2, where a=2k.

$2b^2=4k^2$

$b^2=2k^2$

So now b must be even but in our assumption, a and b are reduced to lowest terms, so both cannot be even. Therefore our assumption is false and $\sqrt{2}$ is not rational.

Is x=0.333333... rational?

$10x = 3.33333$

$-x = 0.33333$

9x=3

x=1/3

Is x=0.14141414... rational?

100x=14.1414141414...

-- x= 0.1414141414...

99x=14

x=14/99

It should be clear from the above examples that any repeating decimal is rational. The idea that $\sqrt{2}$ could not be expressed as a rational number caused a major crisis in the world of fifth century BC Greece.

Other Interesting Problems: Series

Let's say that S=1/2 + 1/4 + 1/8 + 1/16 + 1/32 + ... As we add this series term by term, we are getting closer and closer to 1. Here is an important new idea: the *limit* of S=1.

S=0.9999999... so S DOES NOT equal 1, but the limit of S is 1.

Zeno's famous paradox involves the legendary race between Achilles and the turtle. Achilles runs ten times as fast as a turtle. The two are involved in a 100 yard race where the turtle is given a 90 yard lead. Zeno argues that Achilles never catches the turtle but any form of elementary common sense clearly indicates that Achilles will overtake the turtle.

Here is Zeno's argument:

The turtle has a 90 yard lead as the race begins.

By the time Achilles runs to the 90 yard line, the turtle is now only 9 yards ahead.

By the time Achilles runs another 9 yards, the turtle is now only 0.9 yards ahead

By the time Achilles runs another 0.9 yards, the turtle is now only 0.09 yards ahead

And so on. So Achilles never catches the turtle. Everyone knows that after 10 seconds, Achilles would have caught up to the turtle, so where is the contradiction? The idea of limits is crucial.

Zeno's Paradox has a special place in the history of math. Let's take another look at the Zeno Paradox. If you want to walk out of a room, first you must walk halfway to the door. Then you must walk half the distance of what is left, and then half of that distance, and so on, so that you never get out of the room. Your progress looks something like this:

You have walked 1/2 of the distance to the door

You have walked 3/4 of the distance to the door

You have walked 7/8 of the distance to the door

You have walked 15/16 of the distance to the door

On the one hand, you know you will reach the door, but on the other hand, in mathematics it's not always this simple. The idea of the term *limit* is important.

The Harmonic Series

$S= 1 + 1/2 + 1/3 + 1/4 + 1/5 + 1/6 + 1/7 + \ldots$

This is the harmonic series, and I asked April to show if it did have a limit. We saw above that $S= 1/2 + 1/4 + 1/8 + 1/16 + \ldots$ has a limit of 1. If you add up the first million terms of the harmonic series, you get approximately 12. The series ultimately goes off to infinity and therefore has no limit. I wanted both April and you the reader to prove this. Here is some help:

$1/3 + 1/4 > 1/4 + 1/4 = 1/2$

$1/5 + 1/6 + 1/7 + 1/8 > 1/8 + 1/8 + 1/8 +1/8 = 1/2$

$1/9 + 1/10 + 1/11 + 1/12 + \ldots + 1/16 > 8/16 = 1/2$

$1/17 + 1/18 + 1/19 + \ldots + 1/32 > 16/32 = 1/2$

Do you see what's going on? You keep getting 1/2's forever and ever, and so the sum goes onto infinity. It is a very special series because it moves very, very slowly to infinity. You have to go out 250 million terms to get to 20, and 1.5×10^{43} terms to get to 100.

A Star is Born

In 1728, determining the value of S for the equation $S = 1 + 1/2^2 + 1/3^2 + 1/4^2 + 1/5^2 + \ldots$ was one of the leading problems in math. The mathematician Daniel Bernoulli proposed the problem to one of his students, Leonhard Euler. Bernoulli was convinced the solution was very close to 1.643. Euler astounded the mathematical world with his solution: $S = \pi^2/6$

How did π get into the act? Euler was recognized as one of the two math giants in the 18th century. The other giant was Carl Gauss, who were the two most important geniuses in the history of math.

I urged April to study the history of π, e, and i. These topics are very interesting and important branches of mathematics. Due to the limitations of how many topics I can present in this book, I decided not to include this material. But I must point out there is one remarkable equation called the Euler Identity, which combines 5 of the main characters in math (*e*, pi, *i*, 0, and 1) into one equation.

$e^{\pi i} + 1 = 0$

Do not ever forget this equation.

Number Theory

The very influential Pythagorean Brotherhood were very enamored with numbers. The Greeks called 6 a perfect number because 6 is divisible by 1, 2, and 3, and 1 + 2 + 3 = 6. Another perfect number is 28, which is divisible by 1, 2, 4, 7, and 14, and 1 + 2 + 4 + 7 + 14 = 28. In general, perfect numbers are of the form $2^{(k-1)}(2^k-1)$. The first five perfect numbers are 6, 28, 496, 8128, and 33,550,336.

These types of problems are in a field of math called number theory, a very enjoyable but not very practical field. The main players in number theory were Gauss and Euler.

By the way, I was born April 28, 1936. April is the fourth month, and four is a perfect square. The Greeks called 28 a perfect number, and 1936 is a perfect square (44x44=1936). Only a math teacher would take such a delight in their birthday.

Another simple number theory problem--and a very practical one--is called casting out nines. If the sum of the digits of a number is divisible by 9, then the number is divisible by 9. Take the number 756; the sum of its digitis is 7 + 5 + 6 = 18. Since 18 is divisible by 9, 756 is divisible by 9. Here is a look at the problem.

$756 = 7\times100 + 5\times10 + 6$

$=7(99+1) + 5(9+1) + 6$

$=7\times99 + 7 + 5\times99 + 5 + 6$

Now $7\times99 + 5\times9$ can be represented as 9T, so that $756 = 9T + 7 + 5 + 6$. Since 9T is divisible by 9, everything depends on $7 + 5 + 6 = 18$, which is divisible by 9.

Now consider a three-digit number abc. Prove if a+b+c is divisible by 9, then the number abc is divisible by 9. The proof:

$$\begin{aligned} abc &= 100a + 10b + c \\ &= a(99+1) + b(9+1) + c \\ &= 99a + a + 9b + b + c \end{aligned}$$

Now $99a + 9b = 9T$, so we have $abc = 9T + a + b + c$. Since 9T is divisible by 9, everything depends on $a + b + c$. So if $a + b + c$ is divisible by 9, then the whole number is divisible by 9.

Try and prove for the four digit number abcd that if a+b+c+d is divisible by 9, then the number abcd is divisible by 9.

Re-write the four digit number abcd as:

$$1000a + 100b + 10c + d$$

$$= 999a + a + 99b + b + 9c + c + d$$

Group $999a + 99b + 9c$ together as 9T, which gives us

$$9T + a + b + c + d.$$

Since 9T is divisible by 9, we know that if a+b+c+d is divisible by 9, then the four digit number is divisible by 9.

A little more number theory

The Fundamental Law of Arithmetic states that every number can be expressed as a product of primes.

6=2x3

$12=2^2x3$

$60=2^2x3x5$

$120=2^3x3x5$

Try these: find prime factorization of 1) 28; 2) 100; 3) 23; 4) 32; 5) 365.

Answers are: 1) $28=2^2x7$; 2) $100=2^2x5^2$; 3) 23=23; 4) $32=2^5$; 5) $365=5x7^3$

Now try and find prime factorization of every number from 1 to 20.

Here are a few problems for you to try:

1) Find the smallest number divisible by all the numbers 1-6.

You need to multiply $5x3x2^2 = 60$

2) Find the smallest number divisible by all the numbers from 1 to 7. You must now include the prime 7 in the answer to question 1, so the answer is 60x7 = 420

3) Find the smallest number divisible by every number from 1 to 10.

The smallest number divisible by numbers 1-7 is 420. Then, we can rewrite $8=2^3$; $9=3^2$; 10=2x5.

We have already included 2^2 with the number 4, so to accommodate 8, we only need to multiply by 2.

420x2=840

We don't have a 3^2, but we do have a 3, so to accomodate 9, we need to multiply by 3.

$840x3=2520=2^3x3^2x5x7$

4) Find the smallest number divisible by every number from 1 to 14.

Answer: $2^3x3^2x5x7x11x13=360,360$

5) Find the smallest number divisible by every number 1-20.

Answer: 360,360x2x17x19 = 232,792,560

Do you see why we had to multiply by 2 to accommodate 16?

Diophantine Equations

Diophantine equations involve equations where we have more unknowns than equations. Let me illustrate this with an interesting problem.

PROBLEM #1: With $1,000, a rancher buys steers at $25 each and cows at $26 each. How many steers and cows did the rancher buy?

Let x=number of steers

y=number of cows

$25x + 26y = 1000$

$25x + 25y + y = 25(40)$

$y=25(40) - 25x - 25y$

$y=25p$, where p is a whole number.

Let $p=1$; $y=25$ and $x=14$

Let $p=2$; $y=50$ and $x=-12$

So the only solution is that the rancher bought 14 steers and 25 cows.

PROBLEM #2: A man spends $1.01 on 2 types of stamps, 10 cent stamps and 3 cent stamps. How many of each does he buy?

Let x=number of 10 cent stamps

y=number of 3 cent stamps

$10x + 3y = 101$

$9x + x +3y = 3(33) + 2$

$x = 3(33) = 9x - 3y + 2$

$x=3p + 2$

If $p=0$, $x=2$, $y=27$

$p=1$, $x=5$, $y=17$

$p=2$, $x=8$, $y=7$

$p=3$, $x=11$, $y=-3$

This solution, then has three possible solutions: $p=0$, $p=1$, and $p=2$.

PROBLEM #3: $2x + 3y = 10$

a) find all whole number solutions for x and y

b) find all positive solutions for x and y

It is easy to show one solution, when $x=5$ and $y=0$. But now comes a key step.

A general solution is $x=5-3T$ and $y=2T$

To prove this is right:

$$2x + 3y = 10$$

$$2(5-3T) + 3(2T) = 10$$

$$10 - 6T + 6T = 10$$

So the solution is correct.

When $t=0$ $x=5$ $y=0$

$t=1$ $x=2$ $y=2$

$t=2$ $x=-1$ $y=2$

$t=3$ $x=-4$ $y=8$

etc.

But for part b, the positive solutions are (5, 0) and (2, 2)

PROBLEM #4: Find all positive whole number solutions to $7x + 3y = 41$.

Try it yourself before you look at the solution.

$7x + 3y = 41$

$6x + x + 3y = 3(13) + 2$

$x=3P + 2$, where P is a whole number

If $p=0 \quad x=2 \quad y=9$

$p=1 \quad x=5 \quad y=2$

$p=2 \quad x=8 \quad y=-5$

Since we want only positive solutions, the solution set must be (2, 9) and (5, 2).

The general solution is: $x=2 + 3P$ and $y=9 - 7P$

PROBLEM #5: A farmer buys 100 animals for $100. Dogs are $10 each, cats are $3 each, and birds are $0.50 each. List all the possible solutions.

Let x=number of dogs, y=number of cats, and z=number of birds.

$x + y + z = 100$

$10x + 3y + 0.5z = 100$

$20x + 6y + z = 200$

$- \quad x + y + z = 100$

$19x + 5y = 100$

$20x - x + 5y = 5(20)$

$5P = x$

If $p=0 \quad x=0 \quad y=20 \quad z=80$

$p=1 \quad x=5 \quad y=1 \quad z=94$

$p=2 \quad x=10 \quad y=-14$ ***Impossible***

The two possible answers are p=0 and p=1.

PROBLEM #6: A woman wants to buy 100 copies of Math Fun for $100. Each hardcover book costs $5, each softcover costs $2, and each used copy costs $0.25. How many of each did she buy? Before you look at the solution, try it for yourself.

Let x=number of hardcovers, y=number of softcovers, and z=number of used books.

$x + y + z = 100$

$5x + 2y + 0.25z = 100$

$20x + 8y + z = 400$

$- \; x + y + z = 100$

$19x + 7y = 300$

$14x + 5x + 7y = 7(42) + 6$

$5x = 7P + 6$, where P is a whole number

If p=2 x=4 y=32 z=64

p=7 x=11 y=13 z=76

These are the only possible solutions.

PROBLEM #7 (A BRAIN TWISTER): Now comes problem #9 from Chapter 3. A woman wants to buy 100 presents for $100. Pens are 10 for $1, Irving Berlin CDs are $2 each, and this Math Fun book is $5 each. How many of each does she buy?

Let x=number of pens, y=number of CDs, and z=number of books. Try it yourself before you look at the solution.

$x + y + z = 100$

$0.10x + 2y + 5z = 100$

$x + 20y + 50z = 1000$

$-x + y + z = 100$

$19y + 49z = 900$

$19y + 38z + 11z = 19(47) + 7$

$11z = 19k + 7$, when k is a whole number

$11z = 22k - 3k + 7$

$11p + 7 = 3k$

If p=1, k=6, z=11. This means that x=70 and y=19 when z=11 (70 pens, 19 CDs, and 11 books).

Since I have drifted away from math problems, let me go one step further and (at Aprils urging) relate the story of Moe Berg (third string catcher for the Boston Red Sox). Moe Berg was a brilliant student at Princeton. He was extremely talented in learning new languages. He also was the best baseball player in the school's history. He had a so-so career in baseball but in 1934, the All-Star American Players were invited to Japan to play the Japanese baseball teams. The intelligence agency of the U.S. Government was increasingly concerned about a possible war with Japan. They enlisted the services of Moe Berg to learn Japanese and take pictures of their instillations. So they got Moe Berg on the 1934 All Star Team to spy on Japan. The Japanese would never suspect one of the All Star Players to be a spy. The baseball world was shocked when the third string catcher of the Red Sox was on the All Star Team. On the lengthy boat trip over, Moe Berg was learning Japanese while Babe Ruth and his pals were drinking and running after women. Well, Moe Berg was very successful in getting valuable pictures of Tokyos instillations which proved very helpful 7 years latter when we were at war with Japan.

We now introduce Werner Heisenberg considered one of the two best mathematicians and atomic scientist in the world. The US State Department was very concerned with the possibility of German advancement in an atomic bomb project. When Germany overran

Denmark, Werner Heisenberg visited his old mentor and teacher, Neils Bohr. In the conversation, Werner said the next war would end with an atomic bomb. Neils Bohr escaped to Britain in 1941 and passed on this information with the added note that Werner Heisenberg was the leading atomic physicist in the world and if Germany would make any progress with an atomic bomb Werner Heisenberg would undoubtedly be the central figure in such a pursuit. The U.S. State Department was considering sending in paratroopers into Germany to bill Werner. The plan did not materialize but in 1944 word got back to the British that Werner Heisenberg was to give a lecture in Switzerland. Moe Berg was recruited to learn the German dialect and attend the conference (with a gun) with orders to kill Werner if there were any indication that Germany was making progress in making an atomic weapon. At the lecture Werner pointed to Moe Berg and said he has never seen a student take so many notes. The University invited Werner for a dinner in his honor. Werner said he would be pleased to attend provided no politics were brought up. Moe Berg somehow got an invitation to this dinner and decided he would wait until after the dinner to make up his mind if he were to kill Werner. There he is, Moe Berg (gun in pocket) at the dinner for Werner. Werner saves his own life by making the statement he doesn't think Germany can win this war. Moe concluded the Germans were not making any progress in the making of an atomic bomb. Moe Berg walks with Werner Heisenberg and engages Werner in conversation and it became clear to him that Germany was not pursuing a path to the making of a bomb. Werner's life was spared and Moe Berg went back to Washington with his report.

Back to Math

April and the other club members asked me to introduce them to calculus. Before I did that, I wanted them to consider the following problem:

If $a = b$,

then $a^2 = ab$

$a^2 - b^2 = ab - b^2$

$(a + b)(a - b) = b(a - b)$

$a + b = b$

$2b = b$

$2=1$

What went wrong? I told them it was important for them to find the error before we move on to calculus. They looked at the problem again and again and couldn't find the error. I finally gave in and pointed it out.

If $a = b$, **OK**

then $a^2 = ab$ **OK**

$a^2 - b^2 = ab - b^2$ **OK**

$(a + b)(a - b) = b(a - b)$ **OK**

Now comes the error: $a - b = 0$, and you are not allowed to divide by zero. Therefore, if you divide each side by $(a - b)$, you are dividing by zero.

Another example: $5x = x$. $x=0$ is the only solution, and you can't divide each side by x, because then you would have $5=1$.

Introduction to calculus

At April's request, she asked to be introduced to calculus. Consider $y=x^2$ and we want to find the slope of the line at x=3.

When x=3.1 y=9.61

x=3 y=9

A reasonable thing to do is:

slope = $\frac{\Delta y}{\Delta x}$ = (9.61 - 9)/(3.1 - 3) = 0.61/0.1

The slope is very close to 6.1.

Now consider:

x=3.01 y=9.0601

x=3 y=9

Now let us find this slope.

slope= $\frac{\Delta y}{\Delta x}$ = (9.0601 - 9)/(3.01 - 3) = 0.0601/0.01

slope=6.01.

6.01 is even a better approximation of the slope at x=3 than 6.1. Notice that we never divided by zero.

An even better approximation is:

x=3.001 y=9.006001

x=3 y=9

slope= $\frac{\Delta y}{\Delta x}$ = 0.006001/0.001 = 6.001

It does look like the slope at x=3 has a limit of 6.

Now for a very bold step. Let h be very, very small.

$x=3+h \quad y=9+6h+h^2$

$x=3 \qquad\qquad y=9$

$$\text{slope}=\frac{\Delta y}{\Delta x}=\frac{9+6h+h^2-9}{3+h-3}$$

$$\frac{\Delta y}{\Delta x}=\frac{6h+h^2}{h}$$

Now comes a key point: h is very small, but not zero. So we can divide by h.

$$\frac{\Delta y}{\Delta x}=6+h$$

Since h is very, very small, the limit of the slope is 6 when x=3.

On planet earth, $s=16t^2$, which is known as the Galileo Formula.

t	0	1	2	3	4	5
s	0	16	64	144	256	400

Therefore, in 1 second, an object falls 16 feet.

In 2 seconds, an object falls 64 feet.

Let's find the velocity of the object at t=2.

Let t=2.1 s=70.56

t=2 s=64

Velocity = change in distance/change in time

from t=2.1 to t=2

change in distance/change in time = (70.56 - 64)/(2.1-2) = 6.56/0.1 = 65.6

The approximate velocity is 65.6 ft/sec.

Now let t=2.01 s=64.6416

$t=2$ $s=64$

Velocity = change in distance/change in time = (64.6416-64)/2.01-2)

= 0.6416/0.01 = 64.16

This is an even better approximation of the velocity of an object near $t=2$.

Let $t=2.001$ $s=64.06402$

$t=2$ $s=64$

Velocity = change in distance/change in time = 0.06402/0.001 = 64.02

Notice that it looks like the limit is 64 ft/sec. Now let's take a bold step: let h be very, very small but not 0.

$t=2+h$ $s=16(4+4h+h^2)$

$t=2$ $s=64$

Velocity = change in distance/change in time = $(64+64h+16h^2-64)/(2+h-2)$

$= (64h + 16h^2)/h$

Now h is small but not zero, so you can still divide by h.

Velocity = 64 + 16h. As h gets very, very small, the limit of the velocity is 64 ft/sec.

Let's go back to $y=x^2$. Can we find a general equation for the slope?

pt 1=x+h $y=(x+h)^2$

pt 2=x $y=x^2$

Slope $= \frac{\Delta y}{\Delta x} = (x^2+2xh+h^2-x^2)/(x+h-x) = (2xh + h^2)/h$

If we divide by h, we get slope = 2x+h. As h gets smaller and smaller, the limit of the slope becomes 2x.

At x=3, slope=6

At x=4, slope=8

The key point is that we never divided by zero, but we made h very, very small.

How did that go?

END OF CHAPTER V SURPRISE:

My ten favorite musicals (in order of preference)

1) Showboat
2) Carousel
3) Fiddler on the Roof
4) Oklahoma
5) West Side Story
6) South Pacific
7) My Fair Lady
8) Annie, Get Your Gun
9) Sound of Music
10) Phantom of the Opera

My list (in order of preference) of the five most outstanding songs from musicals:

1) "Ah! Sweet Mystery of Life" by Victor Herbert from "Naughty Mariette" (1910)
2) "Love Changes Everything" by Andrew Lloyd Weber from "Aspects of Love" (1977)
3) "Ol' Man River" by Kerns and Hammerstein from "Show Boat" (1927)
4) "September Song" by Kurt Weil from "Knickerbocker Holiday" (1938)
5) "If I Love You" by Rogers and Hammerstein from "Carousel" (1945)

CHAPTER VI:

TWENTY MATH CHALLENGES

1. A sphere with a radius of 1 inch is placed in a 2" cube. In another 2" cube, we place 1000 spheres of radius=0.1". Which cube has more empty space?
2. A ball is dropped from a height of 1 mile. Each time it bounces, it loses half its height. find the limit of total distance traveled in 1 year.
3. Find the smallest number that is divisible by every number from 2 to 10. Then find the smallest number that is divisible by every number from 2 to 13.
4. A box is 6' x 8' x 10'. find the longest pole you can fit in the box.
5. Frank takes Betty on a date and buys her French fries. Betty asks him how many equilateral triangles he can make with 6 fries of the same length.
6. Find three different numbers such that the sum of their recipricals is 1. Can you find four such numbers?
7. The problem has some interesting applications. The sides of a triangles are (u^2+v^2), (u^2-v^2), and 2uv. The (u^2+v^2) side is the longest. Prove that this is a right triangle.
8. 1+2=3; 3+5=8; 7+5=12; 12+1=1; 7+6=1; 11+11=0; 7+8=? What is going on?
9. Use a calculator for this problem: A parachute company advertises their parachutes work 99% of the time. a) What are the odds of 16 successful jumps? b) After how many jumps would the odds be 1/2?
10. What are the odds of a person with a 0.400 batting average (who gets four times at bat in each game) getting at least 1 hit in eight consecutive games.

11. This is a very interesting problem. In triangle ABC, $\angle A=72^{o}$, $\angle B=72^{o}$, and leg AC=1. Find AB. *Hint: bisect angle A.*

12. A stone is dropped into a well, and 7.7 seconds later we hear the sound of the stone striking water. Assume that the stone falls $s=16t^2$ and the velocity of sound is 1,120 ft/second. Write an equation to find the depth of the well and approximate the answer.

13. The Monkey and the Coconuts. The short story titled "Coconuts," by Ben Ames, appeared in the *Saturday Evening Post* on October 9, 1926. The story tells about 5 men and a monkey who were shipwrecked on an island. they spent the first day gathering coconuts. During the night, one man woke up and decided to take his share of the coconuts. He divided them into five piles. One coconut was leftover, so he gave it to the monkey, then hid his share, and went back to sleep. Soon, a second man woke up and did the same thing. The third, fourth, and fifth man followed the same procedure. the next morning after they woke up, they divided the remaining coconuts into five equal shares. This time no coconuts were left over. What is the smallest number of coconuts in the original pile? This problem is one of the most famous puzzles- it gets the prize as being the most worked on and least solved puzzle in history. *Hint: the solution takes on the form 5^k-4.*

Before you try this problem, please first try this easier problem:

Five men and a monkey were shipwrecked on an island. They spent the first day gathering coconuts. During the night, one man worke up and decided to take his share of the coconuts. He divided them into five piles, his his share, and went back to sleep. Soon a second man woke up and did the same thing. The third and fourth man followed the same procedure. The fifth man (the honest one) slept through the whole night. When they woke up, they divided the pile into five equal shares, with one coconut leftover for the monkey. What is the smallest number of coconuts in the original pile? *Hint: the answer is of the form* 5^k.

For 250 years, the *Saturday Evening Post* survived. The outstanding magazine started by Ben Franklin folded in 1969. Their covers and some of the articles will always be a part of the fabric of America.

14. You have 12' of rope. Find the maximum area you can enclose in the shape of a) a 4-sided figure, b) a semicircle, c) no restrictions.

15. In a 3-digit number abc, prove if a + b + c is divisible by 9, then the number is divisible by 9.

16. Write the number 100 base 10 in a) base 2 and b) base 3

17. 1% of the population has disease C. There is a test for C which is 99% accurate. A woman takes the test and it comes out positive. She is very upset and sad but her husband (a retired math teacher) tells her the odds are only $\frac{1}{2}$ that she has C. The wife said that's incredible but of course he was right. Explain. Hint: Take 10,000 people and see what happens.

18. Prove $S = \frac{1}{2} + \frac{1}{3} + \frac{1}{4} + \frac{1}{5} + \frac{1}{6} \ldots infinite$

Hint : $\frac{1}{3} + \frac{1}{4} > \frac{1}{2}$ $\frac{1}{5} + \frac{1}{6} + \frac{1}{7} + \frac{1}{8} > \frac{1}{2}$

$\frac{1}{9} + \frac{1}{10} + \frac{1}{11} + \frac{1}{12} + \frac{1}{13} + \frac{1}{14} + \frac{1}{15} + \frac{1}{16} > \frac{1}{2}$

19. Another hard problem. A 36' stick is cut at random in 2 places forming 3 pieces. What is the probability of forming a triangle with the 3 pieces?

20. THIS IS MY FAVORITE PROBLEM! I have given it to some of the best math people and they cannot solve it without a hint. Give this problem only to your most brilliant friends. Albert Einstein goes into a Seven/Eleven store and buys 4 items. The clerk (his best student) tells Albert the product of the 4 items is \$7.11. Albert tells the clerk you should add, not multiply. The clerk answers, "May I call you Albert"? Albert says, "Yes, but how could you make such a mistake"? The clerk answers, "Albert, I added up the total and it is in fact \$7.11. I multiplied and also got \$7.11". Albert said, "I am changing y our grade from A to A^2".

What are the prices of the 4 items? Hint: \$7.11 = 9x\$.79

.79=prime

Answers to the above questions:

1. The same: $8-(4/3)\pi$

2. $S = 1 + 1/2 + 1/2 + 1/4 + 1/4 + 1/8 + 1/8 + 1/16 + \ldots$

   $S = 2 + 1/2 + 1/4 + 1/8 + 1/16 + \ldots = 3$

3. a) 2520; b) 2520 x 11 x 13 = 360,360

4. Use the Pythagorean Theorem twice: $\sqrt{200}$

5. Think 3D: 4

6. a) 1/2 + 1/3 + 1/6; b) 1/6 = 1/9 + 1/18; 1/2 + 1/3 + 1/9 + 1/18

7. $(u^2-v^2)^2 + (2uv)^2 = (u^2+v^2)^2$; $u^4 - 2u^2v^2 + v^4 + 4u^2v^2 = u^4 + 2u^2v^2 + v^4$ Thus it is true.

8. Clock arithmetic: 7 + 8 = 3

9. a) $(0.99)^{16}$= about 85%; b) $0.5(0.99)^x$, so x= about 69 or 70; c) 1 jump

10. To get at least one hit, he must avoid NNNN= $(3/5)^4 = 81/625$ = 13%; So 87% of the time, he will get at least 1 hit. $(0.87)^8$ = about 33%

11. You get similar and isosceles triangles: $(1-x)/x = x/1$, $x = (-1 + \text{sqrt}5)/2$

12. $16t^2 = 1120\,(7.7-t)$, t is approx. 7. Distance = $16t^2 = 16 \times 49 = 784$; $1120(0.7) = 784$

13. a) $5^5-1 = 3121$; b) $5^4=625$

14. square=9; semi-circle is $\pi r + 2r = 12$, $r=\frac{12}{\pi+2}$; $A=\frac{1}{2}\pi r^2=\frac{1}{2}\pi\left(\frac{12}{\pi+2}\right)^2$, which is about 8.6.

    $2\pi r=12$, $r=6/\pi$, $A=\pi r^2=36/\pi$

15. abc = 100a + 10b + c = 99a + a + 9b + b + c = 9T + a + b + c. Since 9T is divisible by 9, the number abc is divisible by 9.

16. Base 2: 110010, 64+32+4=100; Base 3: 1021, 81 + 2x9 + 1 = 100

17. See problem #3 in Chapter IV

18. This series contains an endless sequence of 1/2, so it equals infinity

19. Since the sum of the two shortest sides of a triangle must be greater than the longest side, the longest piece is greater than 12 but less than 18. The longest piece is 24" or less, so only 12" to 18" is acceptable to make a triangle. 6:24 = 1:4

20. 711 is divisible by 9 ($711 = 9 \times 79$). 79 is a prime number, so the only reasonable answer for one of the items is $2.37 or $3.16. Intuition tells me that $3.16 is the answer. The four items were priced at $3.16, $1.50, $1.25, and $1.20.

Ten more problems for the eager beavers.

1. At 2:15, the hour hand and the minute hand of a clock form an angle of what?

2. Successive discounts of 10% and 20% are equal to one discount of what?

3. A 25' ladder is 7 feet from a wall. If the top of the ladder slips 4', then the foot of the ladder will slide how many feet?

4. With the use of three weights (1lb, 3lb, and 9lb), how many objects of different weights can be weighed on a balance scale?

5. A total of 10 handshakes was exchanged at a party of polite people. How many people were at the party?

6. With $1000, a rancher buys steers at $25 each and cows at $26 each. How many steers and cows did he buy?

7. A girls' camp is 300' from a straight road. On this road is a boys' camp that is 500' from the girls' camp. A canteen is built on the road exactly the same distance from each camp. How far is it from each camp?

8. Find all scalene triangles having integral legnth sides and a perimeter of 12 or less.

9. The medians of a right triangle which are drawn from the vertices of the acute angles are 5 and sqrt 40. Find the value of a hypotenuse.

10. Abe, Bob, and Sue started on a 100-mile journey. Bob and sue went by car at 25 mph, while Abe walked at 5mph. After a certain distance, Bob got out of the car and walked at 5mph while Sue went back to get Abe. They all arrived at their destination at the same time. How many hours was the whole trip?

Answers

1. In 15 minutes, the hour hand moves 1/4 of 30°=7.5. So 30-7.5=22.5

2. 28%

3. Right triangles of 7, 24, 25 and 15, 20, 25. The ladder moved 8 feet.

4. All weights from 1-13 lbs.

5. Five people. The problem is very easy if you know about permutations and combinations.

6. Do you remember this problem from Chapter V? 14 steers and 25 cows.

7. Right triangle of 300, 400, 500. $300^2 + (400-x)^2 = x^2$. x=312.5

8. (5, 4, 3); (5, 4, 2); (4, 3, 2). Three possibilities

9. $(a/2)^2 + b^2 = 25$; $a^2 + (b/2)^2 = 40$. a=6, b=4, hypotenuse= sqrt52

10. Let t_1=time when Bob got out of the car. Let t_2=time Sue goes back to get Abe (time Sue picks up Abe?). Let t_3=rest of the trip.

$$5(t_1 + t_2) = 25t_1 - 25t_2; 6t_2=4t_1.$$

$$t_3= (100-(25t_1-25t_2))/25 = (100-(25t_1+5t_2))/5$$

t_1=3, t_2=2, t_3=3. The trip took a total of 8 hours

CHAPTER VI SURPRISE

. I hope you had a good ride!

Thanks for reading,
Werner

Summary

This book is designed to be a fun learning experience for students and adults who appreciate the joy of math. I am a retired math teacher from the Bronx High School of Science, and I know a thing or two about teaching and making math interesting. The book contains a variety of puzzles, projects, and discussions related to math. Each chapter also contains delightful surprises to appeal to the reader of all skill levels. You are in for an unforgettable adventure.

About the Author

Werner Weingartner is a retired math teacher from the Bronx High School of Science. He produced, wrote, and directed the movie "1+1=X." He originated the Weingartner Global Initiative, which is an ongoing project at the College of William and Mary in Williamsburg, VA.